KING EDWARD I

6024 PRESERVATION SOCIETY LTD

Treasurer, Capt
The Mill, Padbury,
Buckingham MK18 2BL.
Please enclose a SAE.

give people pleasure by publication of a number of photographs picturing the locomotive running in service days and then showing the contrast between its present condition and how it stood between 1963 and 1973. The book was not written to break new ground. Almost all of the information contained in this book has been brought to the attention of the railway enthusiast public or to members of the "6024 Society" before. The opinions about the development of railway preservation since 1962 are probably a mixture of views held by others past and present.

At the outset it seemed doubtful that we would have sufficient pictures to cover the period of the locomotive's ordinary service running between 1930 and 1962.

Nothing could have been farther from the truth! The engine is one which has had the good fortune to be recorded time and again; it was photographed several times brand-new outside the works, and we know the name of the driver on the first of its trips to Plymouth.

In 1932 Joe Traxler made a silent film from the footplate of 6024. This can be found on Stirling Video's excellent production.

In the 1950s and 1960s Kenneth Leech and Dick Blenkinsop, amongst others, saw 6024 often when taking photographs; and Gavin Morrison found the engine at Cardiff on the week of its withdrawal from service in 1962.

Possibly no other "King" has been so well recorded save 6000 which was recorded from the very first with proper G.W.R. pride!

Additionally, the Society still has to face heavy maintenance costs because the locomotive is, of course, one of the largest in the country. We hope to be able to raise a certain amount of funds by the sale of the book, particularly to enable us to retyre the locomotive. That is a job which we shall face within the next two years and we would like to be ready for it because otherwise the value of the initial seven-year boiler certificate is going to be much attenuated.

It is a satisfactory and satisfying time to be able to write such a book. Satisfying to the writer in his four years' membership of the Society's working parties to have watched a set of wheels and frames gradually assemble itself into a working locomotive by a dint of high standards of painstaking workmanship. Satisfactory because the writer, amongst many others, saw the locomotive not in running days but in the scrapyard when it had been shorn of most of its non-ferrous parts and was in a dilapidated state and the contrast has brought it to a condition on a par with other mainline locomotives which have run in recent years.

Readers will recognise the birth of the Society, the rescue of the locomotive from the scrapyard and the events involving the Society between 1973 and 1989 as a paradigm of events facing many other railway preservation societies. The 6024 preservation project had its origins at a time when few locomotives rescued from Barry scrapyard had been returned to running condition and has covered the period to a stage when the scrapyard has been emptied of locomotives. The experiences of raising the initial funds to remove the locomotive and find a site for its restoration and then the financial and practical setbacks experienced before the machine could run again must be experiences common to many others taking on similar projects. The people who persisted when funds were low must now feel especially gratified and satisfied with themselves for their persistence for so many years before a light appeared at the end of the tunnel. It is worth recording that until about July 1986 work on the locomotive took place out in the open in a very exposed part of the country and to get the frames and other steel parts of the locomotive up to preserved and painted standard was no mean feat. Members of working parties who have attended the site every fortnight throughout the year have probably put in, with travel, the equivalent of between 5 and 7 working weeks in each year.

The wife of one especially long-serving working party and committee member was heard to remark on the steaming day of 2nd February 1989 when a temporary problem arose that she had hoped that events that day would bring to an end the absence of her husband from Sunday lunch for what must have been about ten years on a fortnightly basis apart from holiday breaks!

Acknowledgments

Thanks are due to Dick Riley for not only the permission to reproduce photographs, but also for advice and time given generously and the loan of his article describing a run on 6024's footplate; to Dick Blenkinsop for considerable background information with his photographs; to Kenneth Leech for giving his time to tell me about his pictures as well as permitting their reproduction; to Joe Traxler and Colin Jacks for their articles and consent to their use in the book; to the staff of the library at the National Railway Museum for directing me to further information about the modifications to the "Kings"; to Ian Allan Ltd. for permission to quote from contents of "Modern Railways" and "Trains Illustrated"; to the "Railway Magazine" for permission to reprint and refer to extracts from past issues; to the Bandcroft girls and the Brown family for reading the drafts and suggestions; to Nicola Powell for her typing skills and retyping skills following alterations!

Thanks are also due to both past and present members of the "6024 Preservation Society and Working Party" for their assistance in giving information to me for the book including Rod Thomas, Darryl Polley and the present members of the committee and other members of the Society and Working Parties for reading drafts and much valued advice and corrections; to those who have allowed the reproduction of photographs or slides of the locomotive in accordance with the acknowledgments under each picture contained in the book (except where the owner or author is unknown).

1: Introduction

Railway historians trace the development of the "King" class locomotives from earlier developments and designs of G.J. Churchward who was the Chief Mechanical Engineer of the Great Western Railway from 1902 to 1922.

Churchward's early designs included the "City" class of 4-4-0 locomotives of which "City of Truro" was one. Churchward was, it seems, not a man to rest on his laurels but instead was constantly in search of improvements in economy and performance. In about 1902 Churchward began investigations into a number of potential improvements to GWR steam locomotives including:-

a. The use of tapered boilers

b. Investigating the benefits of use of high-pressure steam by running experimentally three French "De Glehn" compound locomotives

c. The use of longer travel valves

d. Work towards standardisation of locomotive component parts

The purpose of tapered boilers was to spread the heat from the firebox at the back of the boiler as evenly as possible around the front end of the boiler by narrowing the heated area at its point farthest from the source of the heat which was the firebox. This practice derived from American locomotive practice in the late 19th Century. In that period British locomotive manufacturers had also moved to the use of longer boilers and thus longer locomotives. It appears that it had been found that some fireboxes with relatively large grate areas had the effect of passing too much heat to the front of short boiler locomotives and that this heat penetrated through to the smokebox doors which were left glowing at the end of journeys. Sparks were also produced in profusion from the chimneys. This indicated that valuable coal was being wasted. Hard driving on any engine would tend to produce such effects.

Churchward's long fireboxes placed between the frames enabled their heating areas to penetrate deep into the boiler with a partial "element" effect which resulted.

Churchward ran the French locomotives for a number of years but did not adopt the compound system of high and low-pressure cylinders favoured by the French. He believed his valve gear improvements together with superheating steam would adequately compensate for not taking up the savings produced by compounding.

The use of superheater elements in steam locomotives enhanced and complemented the "element" effect referred to above on the steam produced by the boilers.

The larger locomotives with long fireboxes and boilers which Churchward began to design produced more power and were heavier. This in turn led to the need for six and eight-coupled locomotives which he introduced both for freight and passenger work so that the weight would be evenly and reasonably widely distributed over the axles of the locomotive.

Below:
No 6024 "King Edward I" as restored making its debut with the Tyseley–Sratford-upon-Avon workings on 15 April 1990. *[Peter Nash]*

Churchward's conclusions led to the building of the "Saint" class, his first 4-6-0 locomotive design. They were considered to have been the most advanced British locomotives to have been designed and built in this country at the time – 1911. The appearance of the class followed the production of his No.100, a prototype.

They possessed:

a. *Long travel valves which gave the driver greater ability to utilise steam according to the requirements of the locomotive, and use the expansion properties of pressurised steam to its fullest extent*

b. *two "simple" cylinders – no compounding*

c. *superheating of the steam – to extract greater use of the heating surfaces in the boiler by the means of passing back heated steam through the hot flue tubes to heat it a stage further*

d. *high-pressure steam boiler at 225 lb per square inch which was higher than the usual maxima of contemporary locomotives. Their boiler pressures were usually no more than 180 lb per square inch.*

Following Churchward's successful introduction of these locomotives, he introduced the "Star" class. These were more powerful than the "Saints". They had four cylinders, two of which were set between the frames under the smokebox and drove the front pair of driving wheels, the outside pair powering the middle driving wheels. They were designed to haul the heaviest and fastest trains on the line.

On G.J. Churchward's retirement in 1922 he was succeeded in his post by C.B. Collett who continued to design very much in the same mould as G.J. Churchward. In doing so, he laid out the pattern for and set the mould for GWR and Western Region locomotives until 1951 and locomotives to Churchward's and Collett's designs ran until the end of steam in the 1960s.

The "Stars" used the same "No 1" boilers as the "Saints" but with 4 cylinders provided a greater power output. One of the greatest advantages of 4 over 2 cylinders in steam locomotives was found to be in the difference of the effect upon the track. On 25th October 1909, G.J. Churchward wrote a letter to A.V. Goodyear.

In it he summed up some important parts of his philosophy of locomotive design:

"....It is correct that there are great advantages in the 30in stroke in the matter of expansive working. The long stroke in relation to the bore is the only way we know of making the simple engine equal in efficiency to the compound engine. You are correct in assuming that the adoption of the 4-cylinder design was not on account of any dissatisfaction with the 18in by 30in cylinders – the relation of stroke to bore is even greater in the four cylinders than in the two cylinders. The four-cylinder engines are naturally better balanced and have their working parts very much lighter, so making them more suitable for high-speed expresses than would any two-cylinder engines of equivalent power...."

Above:

G.W.R. "Star" class locomotive No 4041 "Prince of Wales" pictured waiting at Exeter St. David's station. The locomotive was introduced to traffic in 1913 and survived into British Railways' service, being withdrawn in May 1951. The loco carried enlarged 15" diameter cylinders, later standard to the class.

One of the fascinating exhibits at the Railway Museum in Swindon is a set of the GWR valve gear for a four-cylinder engine. The construction can be examined in detail, whereas on the locomotives it is very difficult to view the parts as a system.

In 1923, C.B. Collett introduced his "Castle" class with No 4073 "Caerphilly Castle" which is now preserved in the Science Museum in London. The design was an improvement upon, and a development of the "Star" class having a larger boiler and firebox and larger valves and cylinders.

As numbers of the "Castles" were built, their design was further improved. The improvements continued into British Railways days and included the fitting of some locomotives with double chimneys, extended smokeboxes, more superheater elements and mechanical lubricators.

Streamlining was tried on "Castle" No 5005 and "King" No 6014 but in a rather halfhearted manner compared with the comprehensive streamlining adopted by the LMS and LNER in pre-war days. The streamlining was eventually removed from both locomotives apart from the "King"'s wedge-fronted cab and the GWR obviously thought little of the usefulness of such modifications.

2: Competition and improvement

After 1900, competition between companies and cost-cutting to maintain profits were considerable incentives and the GWR and other British railways continued to enlarge the size of locomotives.

In some places companies were competing against each other on lines to the same destinations running along broadly similar routes. They were vying with each other for passenger and freight business.

The increasingly even spread of wealth in the country after the First World War meant that travel and the need to convey goods took place on an ever increasing scale.

Larger locomotives avoided the need to divide trains into two or run one long train with two locomotives double-headed or using a banker in the rear along heavy routes. That enabled the railway to cut down on the consumption of coal and water and also the number of crews needed to run trains.

To the railways a particular advantage to be gained when competing with other railways was to cut timings on routes. The use of larger and more powerful locomotives was the answer.

After the First World War, the GWR was, like other railways, under financial pressure brought about by the effects of the War on the British economy. During the War, stock had been rundown through intensive use with some lowering of servicing standards and this happened too in the Second World War. Galloping inflation also overtook the British economy. It raised the cost of materials and coal and brought Union pressure to increase wages. By 1930 wages and prices combined were 41.91% for the Railway more than in 1913.

Wages alone had increased for the GWR by 62.61% between 1915 and 1930, for the labour cost of engine production (GWR figures).

The cost of materials for the renewal of engines was 41.65% more in 1929 than in 1913, according to GWR records.

So the GWR sought to become more efficient by cutting costs. The GWR had a continuous policy of renewal of engines using serviceable parts of older engines on occasions and replacement with more powerful and efficient locomotives. G.J. Churchward's long-travel valves and superheating of steam and other efficient aspects of boilers were good examples of efforts in this direction.

The railways carried out continuous cost analysis of wages, amounts of coal and water used per mile, considering the value and usefulness of residual scrap values of locomotives and other rolling stock in order to ensure that they retained control over the way in which their resources were used.

In these ways they sought to maximise profits.

After the 1923 grouping of small companies, many locomotives of small constituent companies of the GWR were gradually scrapped or fitted with standard GWR equipment such as boilers when they came to Swindon for overhaul.

The fact that C.B. Collett continued to follow many of the basic design principles of G.J. Churchward, his predecessor, ensured that, together with the work of the successor to C.B. Collett who was F.W. Hawksworth, the last Chief Mechanical Engineer of the railway before Nationalisation, the GWR benefited from nearly half a century of likeminded locomotive designers. F.W. Hawksworth had been one of the senior design staff during the building of the "Kings".

Standardisation and continuity were valuable watchwords to the GWR during this period.

In the 1920s and 1930s another way in which the railway sought to maintain profit was to improve its routes. A number of examples are to be found.

On the London to Birmingham line, a direct route out of Paddington from Old Oak Junction rejoining the longer route via Oxford at Aynho Junction via High Wycombe and Bicester had been built in conjunction with the Great Central Railway. In the 1930s that route was further upgraded with the building of fast and slow up and down lines between Birmingham and Solihull.

These works kept the GWR on level terms with its competitor on the vital route between England's two largest cities. The rival company, the LMS, ran its trains to Birmingham from Euston via Watford and Coventry. Rivalry in Birmingham was such that although a link between the companies' two lines was partly built, it was left unfinished and the partly built viaduct still stands near the centre of Birmingham. The LNWR, the predecessor of the LMS, refused to agree to the completion of the link because it feared the GWR would obtain an advantage with its completion.

Down in the South West the competitor was the Southern Railway on the route between London, Exeter and Plymouth. Both prestige in the running of boat trains to Plymouth and revenue were at stake because the South West was a valuable holiday area. In addition both companies ran to Plymouth and Weymouth with boat trains and import perishable goods to other parts of the country. Between the two World Wars the GWR Ocean Liner Specials ran at the greatest frequency; tenders meeting transatlantic liners and ferrying passengers and cargo to Millbay Station in Plymouth to transfer to London trains. At Weymouth the trains met Channel Island ferries at the adjacent harbour station.

In 1930 the Frome and Westbury "cut-offs" were constructed. These bypasses of two towns on the line served as lines to enable expresses to overtake slower trains and to avoid the slacks imposed on running through these two junction stations.

In July 1906 the GWR had already opened the direct line from Westbury to Cogload Junction near Taunton to cut out the running of trains to the West through Bristol.

By 1925 the GWR directors had the benefit of two years' experience of running their latest and largest express passenger locomotive – the "Castle" class.

Its design had been widely praised. Examples had been run on the LMS and LNER and had been the instigation of valve redesign of the "A1 Pacific" class on the LNER. The Empire Exhibitions of 1924 and 1925 had enabled the GWR to show off the engine with a boast that it was the most powerful express locomotive in the country.

In 1925 the GWR directors issued a statement that they saw no need in the foreseeable future for a larger express engine than the "Castle" class.

Whilst Swindon works continued to build the "Castles" beyond the end of the existence of the Company into 1950, they did not hold to the 1925 view for long however. Their competitive attitude took a hand.

3: The development and building of the "Kings"

In 1925 the Southern Railway facing GWR competition in the south of the country developed and introduced the "Lord Nelson" Class 4-6-0 locomotive. Compared with earlier Southern express classes these engines were built bigger in every way. They were the result of a new design from SR Chief Mechanical Engineer, R.E.L. Maunsell. The boiler was massive and their tractive effort, at 85% of full boiler pressure, was 33,510 lbs.

These locomotives were used on the Southern lines to Southampton, Bournemouth, Weymouth and Exeter.

The introduction of the "Lord Nelson" class meant that the GWR could no longer boast. Their "Castle" class, with a tractive effort at 85% of boiler pressure of 32,160 lbs, was marginally less powerful.

Suddenly it seemed the GWR directors changed their view expressed not long before. The General Manager at the time, Sir Felix Pole, expressed a view that he was not completely happy with some features of the "Castle" class.

The upshot was that the GWR decided not to be outdone by the Southern Railway. Moreover, trains were becoming heavier and speeds were picking up again after the immediate post-War years and the depression of 1929 and the 1930s was not yet in sight.

Not for the first time the GWR considered increasing the size of its largest engines. Previous outcomes had been "The Great Bear" and drawings still exist proposing a "Super Castle" with "No 7" boiler.

Below:
An overall view of the famous Swindon "A" Shop showing no less than 5 "King" class locos of Lot No 243 in course of erection in Spring 1927. A boiler is being lowered into the frames of one of these engines. The unique "King" bogie is seen bottom left in course of assembly; in all, the frames of 6 of this first Lot of "Kings" are visible.
[Crown Copyright – Public Record Office]

C.B. Collett was instructed to design a larger express class than the "Castle". The decision to build a locomotive and the design and construction work was thereafter carried out with speed (unless the 1925 statement referred to above about the adequacy of the existing "Castle" class locomotive was merely a diversion to other railways to hide the plans to build a "King" class locomotive!)

The frames of No 6000, the first of the class "King George V" were laid down in February 1927 at Swindon. From remarks of those who went around the works as outsiders at the time, the building of the locomotive was not publicised. The engine was being run in by July that year. In August 1927 it was shipped to the USA from Cardiff for the Centenary Celebrations of the Baltimore and Ohio Railway.

The building of "King George V" was carried out with great attention to publicity after its completion. The construction was photographed at different stages and the GWR later produced an album of the pictures which can still be seen at the Public Record Office in Kew. The shipping out of the engine itself to America was a publicity coup.

The "Kings" were considerably more powerful than the "Lord Nelson" class by virtue of a higher tractive effort at 85% boiler pressure of 40,300 lbs,

In the 1930s however, the diameter of the cylinders was reduced from sixteen and a quarter inches to sixteen inches slightly reducing that tractive effort. By then though the GWR had achieved their desired effect as to publicity.

The GWR records also show that the runs undertaken by "King George V" in America were meticulously recorded by observers. A GWR driver and fireman travelled with the locomotive and drove it. Even now "King George V" carries on its cab side two medals from the Baltimore and Ohio Railway together with a commemorative bell above the front buffer beam.

The first works "lot" for the building of the first twenty "Kings" was No 243. There was no interval between the building of No 6000 and its subsequent class members up to No 6019. The photographs of the building of 6000 show frames of five other "King" class locomotives with it.

The design of the "King" class was a locomotive of markedly larger dimensions than the "Castles", but even to a layman, of similar style and obviously from the same design stable.

The boiler of six foot diameter was larger than the "Castle" boiler at five foot. The pressure was higher at 250 lbs per square inch and the firebox was bigger at 11'6" in length outside. The engine possessed the same wheel arrangement and the firebox was set between the frames. There is no spare room between the heads of the stays on the outside of the firebox and the inside faces of the frames as those of our Society witnessed when the boiler was replaced on No 6024 after overhaul.

One almost conjured up visions of a giant shoehorn.

The driving wheels at six feet six inches diameter were smaller than the "Castle" driving wheels of six foot eight and a half inches diameter and the "Kings" were longer than the "Castles" and the wheelbase was longer – 29'5" against 27'3". There is markedly more space between the wheels of the "King" as compared with a "Castle".

"Kings" possessed the same arrangements of valve gear as the "Castles" and "Stars". Rocking shafts were used to drive the valves of the outside cylinders. This system can be contrasted with the system of four sets of Walschaerts valve gear adopted by Stanier when he designed his "Princess" Pacific locomotives on the LMS after moving there in 1933.

The "Kings" are also distinguishable from the "Castles" by virtue of the external springs to the front pair of bogie wheels, The springs will not fit inside the bogie frame because they would foul the pair of inside cylinders set under the front of the smokebox if this were done. Initially coil springs on each end of the leaf spring hangers of these front springs were absent.They were added because the locomotives as built displayed some bogie instability. This came to light when No 6003 suffered a bogie derailment at Midgham, Berks in 1927. Another modification to cure this activity was to increase the number of leaves in the rear driving wheel springs.

Above:

On 7th September 1958, the bogie from 6024 "KE1" photographed in Swindon "A" Shop during a heavy intermediate overhaul which lasted from 28.7.1958 to 1959 on and off; the bogie shows the final design modifications including:

i) more leaves and stronger coil springs to the front wheels following the 1927 Midgham derailment and

ii) the strengthening plate welded to the frames after fatigue cracks were found in 1956.

[R Blenkinsop]

Left:

A photograph of "Lord Nelson" class No 863 "Lord Rodney" of the Southern Railway leaving Victoria Station just before the Second World War. It is said that the introduction of this class was the signal for the G.W.R. directors to build an express passenger locomotive larger than the "Castles". *[E. Treacy – Millbrook House Collection]*

The result of building the "Kings" was therefore an increase in tractive effort by about 20% over the "Castles" and brought considerable prestige back to the GWR and its locomotive running department.

The GWR had been obliged to uprate the condition of some of their track and bridges to enable them to run the "Kings" over the main routes. The track had been progressively uprated by the civil engineer in the years before 1927, apparently unknown to Collett! At 22.5 tons the axle loading of the "Kings" is the heaviest for a locomotive on any railway in this country.

With the "Kings" available the GWR took the opportunity to cut down the running times of their main expresses including the "Cornish Riviera Express" from London to Plymouth to exactly four hours. The locomotives were able to take heavy trains over the steep South Devon banks without banking locomotives up to a limit of 360 tons.

Interestingly, soon after introduction of the class, the bore of the cylinders of the "Kings" was reduced from 16.25" to 16" slightly reducing the tractive effort previously claimed for them.

It was over the South Devon banks that the GWR had run certain trials with two members of the "King" class. No 6002 had 16" diameter cylinders and 6005 had 16.25" diameter cylinders. The Locomotive Carriage and Wagon department ran both engines, each with thirteen empty coaches in their train between Plymouth and Newton Abbot over the banks on the 6th November 1927. Particular attention was paid to the running of the locomotives on Hemerdon and Dainton inclines and with No 6005 only the trials were repeated on the 27th November 1927. The stated purpose was to find out whether No 6002 was as strong as the other "King" class locomotives.

On the 29th November 1927 following the trials, Mr F.C. Hall wrote in a memorandum to Mr Collett ...."...the extra power obtained by the slightly larger cylinders is reflected in the superior results obtained on the banks."

Despite this conclusion the cylinders of all the "Kings" were reduced to the slightly smaller diameter.

The "Diary" of No 6024 in the works locomotive record shows the following two entries relating to the cylinders:

"MAR'38 – S,don. Cylinders. Right hand outside renewed"

"NOVEMBER '45 – SWINDON. New cylinders inside only sixteen inches by twenty eight inches"

Another part of the record shows that both outside cylinders were renewed in February 1957.

(The same records show that 6026 "King John" received new inside cylinders on 2nd December 1961, not long before withdrawal in July 1963!)

4: The building of 6024 and its prewar work

Ultimately 30 "Kings" were built at Swindon. No 6024 "King Edward I", the Society's locomotive, was built in June 1930 at a cost of £7,175.

6024 was one of the second batch of "Kings" built at Swindon. The ten locomotives 6020 to 6029 inclusive were built under the works designation of "Lot 267".

The decision to proceed with Lot 267 was taken in 1929. The country was shortly to enter the depression years of the 1930s but the GWR found it still desirable to proceed with the building of the locomotives in 1930. Possibly it was thought that the depression would be shortlived.

The company balanced the construction of Lot 267 and other new locomotives against the withdrawal of some older types which had outlived their usefulness.

The Swindon records show in an extract from minutes of a meeting of the Locomotive Committee at Paddington on the 29th July 1929 that it was agreed to recommend the construction of certain rolling stock on "renewals account".

In particular the building of 100 new locomotives was decided and the details from the minutes of the meeting are reproduced below (the sums of money are the estimated costs of the building of each type):

10	4-6-0 "King Class"	£75,000
20	4-6-0 "Hall Class"	£112,000
30	2-6-2T "51XX Class"	£123,000
10	2-8-0T	£46,000
30	Auto engines of new type	£90,000
TOTAL		**£446,000**

Set against these was the recommended withdrawal of 100 locomotives whose renewal cost was estimated to be £365,000.

The engine and original tender of No 6024 were both built in the first half of 1930. The engine was completed on 30th June 1930.

When built the locomotive was fitted with a standard "No 12" boiler No 4687.

The tender firstly coupled to the loco was No 2552 of the 4,000 gallon type. Ultimately the loco was to haul a number of different tenders in its working life.

The records show 6024 to have been "set to work" (in the words of the records) on the 5th July 1930.

On the 12th January 1932 by the time of its first withdrawal for intermediate heavy repair, 6024 had run 96,552 miles.

The Society is fortunate in having Mr J. Traxler's description of his footplate ride on 6024 on Saturday 23rd November 1932.

"In 1932 I was working as an Assistant Engineer for Messrs W.T. Henley's Telegraph Works. We were engaged on the construction of the 132 K.V. Overhead Electricity Grid System in South Wales. The steel work for the towers constituted a large tonnage, and the Great Western Railway handled the traffic. Through the kind efforts of the South Wales Goods Manager at Cardiff, Mr Cracknell, arrangements were made for me to make the trip on the engine from Paddington to Plymouth and to film the run with my 16mm camera. Up to that time I was the first layman to be granted a footplate pass on a G.W.R. engine, and was highly privileged. This was to be a lifetime's ambition fulfilled.

On Saturday, November 23rd 1932 I walked the length of Platform No 1 at Paddington totting up the weight and composition of the 'Cornish Riviera'. Fourteen vehicles, 480 tons tare, 520 tons full as far as Westbury where two coaches would be slipped for Weymouth leaving 445 tons gross to Taunton. Here the next two coaches would be slipped for Minehead and Ilfracombe respectively, leaving 375 tons full to go forward to Exeter, where a stop would be made and a coach detached for the Torquay line. Our load thereafter to Plymouth would be 9 vehicles, 340 tons full to take unassisted over the Dainton and Rattery inclines.

10.18 a.m. as No 6024 backed onto the train and coupled up. Mr Fred Sheldon, a Swindon locomotive inspector took charge of me and I was introduced to Driver S.Braund, a jovial Devonian and Fireman Rowe, a man in his early thirties. The crew and engine were from Laira shed, Plymouth. Mr Sheldon was wearing the blue Melton overcoat with velvet collar, and black bowler hat customary for Loco Inspectors at that time, while Braund, not liking the regulation Drivers Hat donned a grubby old soft cap. Rowe was attired in the usual overall but to my wonderment wore a stiff 'up and down' white collar. After all, these chaps were the cream of the Top Link. Braund tested the braking system to 25 inches of vacuum, and made some adjustment to the hydrostatic lubricators – then took out his great silver watch and nodded. It was 10.20. He never looked at that watch again all the way to North Road.

Braund then noticed Fred Sheldon tying up the bottoms of his trousers with pieces of string. With a grin and a twinkle in his eye he called across the cab "Wha'be doin' then me old dear? Think we'm in for a rough trip do 'ee". But with the roar from the fire and boiler Fred Sheldon affected not to hear and I quietly put on some cycle clips.

Jim Varcoe the Guard then arrived and said "Mornin' Mr Braund. Nice quiet day for us-no fog. Us've got fourteen on for four-eighty. two to come off at Westbury, two off at Taunton and one at Exeter. Nice comfortable load for 'ee. Should have a good run, eh?"

Braund grinned and replied "Aw 'ullo me old dear. Quiet day sure enough. Yes, we'm aimin' to do our best today-we got company y'see". All the train personnel seemed on good terms, respecting each other for the experts they were in their different jobs, making for smooth working all through.

Fred Sheldon, making himself heard above the roar shouted in my ear – "This engine's done 150,000 miles since last to the shops for general overhaul but she's

Left:
A view taken outside the running shed at Swindon on 2nd July 1930. It depicts a probably unique occasion: the gathering of 7 "King" class locomotives. It took place on the emergence from the works of 3 new locomotives, Nos 6022,6023 and 6024. The engines are, left to right: 6005 King George II
6008 King James II
6017 King Edward IV
– all of these from the first Lot built 1927 – 1928; of the 2nd lot:
6020 King Henry IV
6022 King Edward III
6023 King Edward II
6024 King Edward I
The last three locomotives were brand new and 6020 was built in April 1930. *[Hulton Picture Library]*

Above:
2nd July 1930. In this photo 6024 has been drawn forward for a "portrait". The engine records show she was put to work on 5th July 1930 and was sent to Plymouth Laira shed to take up her duties. The society has it on good authority that Driver Mitchelmore of Plymouth took 6024 down to Laira on the day of delivery: Mr Frank Mitchelmore, his son, is now 84 years of age. He visited Quainton Road Station on one of 6024's steaming days in 1989. *[Hulton Picture Library]*

in good nick, sweet and very free-running. Some say the "Castles" are faster with their larger driving wheels, but these "Kings" have the power to call on. One of them has done 120 on the Test Plant at Swindon so we know they can move. A few years ago one of them on an Express near Newbury had a bogie partly derailed, but no harm done as luck would have it. They cured it by altering the springing – safe as houses now". He went on to explain the G.W. A.T.C. System – the 'Little friend in the corner' of the cab. "Its a fine thing" he enthused "especially in foggy weather. Another G.W.R. invention resulting in better safety and traffic movement". He pointed along the platform "And there's another one. Searchlight signals coming into use, and some also on semaphore signals. A great help". The signal in question was 'off' with its green aspect showing.

Fred Sheldon looked at his watch and said "Well, we should have a good run. There's a P.R. Check for about a mile near Thatcham – probably cost us five minutes in running, but we'll soon get it back. Should put a bit of spice into the trip". Rowe adjusted the live steam injector, leaned out to watch the overflow, closed the valve and also the firehold doors to a mere slit behind which the fire glowed almost white, built up Great Western fashion to a bit above the lower deflector plate. Water level was three quarters up the glass, safety valves roaring with 250 p.s.i. on the gauge. It was almost time to go.

It was quite evident that I was in company with good men whose quiet efficiency brought the train to a state of readiness a bare minute before the start. Ten-Twenty-Nine and a Half. Fred Sheldon and Rowe were looking back along the train, and Braund had his hand on the regulator handle. The last van doors were closed. 6024 was 'blowing off' with a real roar – a long thirty seconds. A whistle shrilled and a green flag waved. Braund pulled the whistle chain, lifted the regulator three quarters open and closed it to the first valve as the first blast left the chimney, then full open and King Edward I took hold of it without the faintest suspicion of a slip with one fifth of a mile of train behind her. Ten thirty dead-on – a fine piece of management.

Filming on the footplate was not easy. It was a dull grey day, and with Kodak pan film stop openings had to be generous. Film speed would be about 13 din. Diaphragm opening about f2.8 for cab interiors – f.8.for general outside shots. But focussing became the greatest problem, needing constant alteration from 2 feet to infinity (no zoom lenses then). Changing the films was tricky too. I had to squat down in the left-hand corner of the cab facing the rear, shielding as best as I could from flying coal dust. Fred Sheldon very thoughtfully assisted me by using his overcoat. This operation had to be performed at Pewsey and near Teignmouth.

As we came out on the straight under the bridge the starting clatter settled down to a concerted rumble, with a slight dither from the side sheets of the cab. Rowe started the exhaust steam injection to supplement the live side, then sprayed water over the coal to keep down dust.

We were on the mile a minute mark by Southall. For years the G.W.R. timetables called for 11 minutes on the first 9.5 miles out, but drivers with full loads never pressed their engines this soon on long runs, so we were the usual 1.5 minutes down at this point. Driver Braund sat on his tip-up seat with all senses tuned to coaxing this temperamental machine into the higher realms of speed. When we reached seventy miles an hour, the rolling motion made a new technique in filming necessary. I had to stand with knees bent and on full tip-toe to damp out the small vibrations which became apparent through the view-finder. Indeed, not until we reached Reading did I really find my 'footplate legs'.

We were doing 69 m.p.h, through Slough with the regulator on the first valve and the cut-off at 16%. A mere whiff of steam was coming from the copper-capped chimney as we watched 6024's great green tapered boiler above the long black rivetted footplating, cutting the wind apart. The arrears of time were a mere half minute now. Rowe's ministrations to the fire were almost continuous, taking care to close the firehole flap plate between each shovelful. There was great dexterity in the use of the shovel – a twelve foot throw to the front of the firebox called for an accurate hump and angle of shovel at the firehole, and Rowe's feet seemed glued to the rolling cab floor.

Then we were through Taplow and over Brunel's flat arch bridge across the Thames into Maidenhead – on time at 80.5 m.p.h. 248 lbs on the gauge, regulator 1st valve in 17%.

Soon we were through Sonning Cutting and approaching the Distants for Reading. Braund closed the regulator, put her in the 45% drifting position and lightly applied the brakes to bring us down to 40 m.p.h. for the slack – then gave her the first valve to maintain reduced speed. Brakes off again and we drifted.

"There's the Junction" Fred Sheldon shouted. "Bristol lines straight on – we fork left for Newbury". The junction was taken smoothly and looking back it was a grand sight to see the long line of chocolate and cream coaches obediently following our 'King'. Braund put her in 30% and full open for a few moments through Reading West, Rowe firing to the back end of the box.

At Southcote Junction the line to Basingstoke diverged to our left as we swung right, wide open to get the train away. A feather of steam from the safety-valves accompanied a 250 lb pressure. The fire, a white hot mass showed that Rowe had prepared for the thirty-two mile climb ahead of us. The signals could be seen at 'clear', came back and swept past us. The curves were wide and banked, and we were rolling round them at 70 m.p.h.

We swung out into a straight to see two long water-troughs glistening between the rails. Rowe had crossed over to the right-hand side of the cab behind Braund. Quickly he twirled the scoop-handle, dropping the scoop into the trough. White spray spumed from beneath the tender and the float-gauge rose rapidly. When Rowe raised the scoop no less than 1,500 gallons of water had rushed up into the tender tank within fifteen seconds – just short of swamping us from the overflow.

A 'Distant' warned us of the 30 m.p.h. P.W.R. Slack as we approached Thatcham. Braund gave a partial brake application and put her into the drifting position and we were down to 30 m.p.h. by the time we passed the Commencement Board. 6024 with the safety valves sizzling crawled forward as though 'shivering with suppressed passion'. The platelayers stood aside to watch us pass and you could see the pride on their faces in one of G.W.R.'s finest engines. The Foreman Ganger waved as we came abreast of the 'T' Board. Braund replied by lifting the regulator full open and the King let out a 45% blast to the sky getting the huge train away in real Great Western style – a thrilling demonstration of Steam Power at its best. Braund settled her down gradually to a quiet purr in 20% at three quarters regulator as we cleared Newbury at 58.5 m.p.h. climbing all the while. We were on the 70 m.p.h. mark through Kintbury with 58.5 miles from London behind us 'on the hour' at 11.30 a.m. Braund's artistry with the King showed itself in the way he maintained 70m.p.h. by using three quarter regulator on the straights, and in order to counteract the drag of the 14 coaches on the curves just bringing up the regulator to full open. Nice judgement. As we swept the curve to Hungerford 27% produced 75m.p.h. Rowe firing continuously 8 or 9 lbs of coal at the time keeping the pressure nicely at sizzling point. The King was happy, responding sensitively to a mere single percentage of cut-off alteration. Braund, with his old black pipe well alight looked across and grinned, then resumed his watch ahead. Fred Sheldon's brief was to see that I did not get in the way or pitch headlong overboard while filming and logging the run.

Approaching Savernake Braund eased her down to 50m.p.h. over Wolfhall Junction, then set her in 15% on the first valve. The gauge was down to 240 lbs – 70 miles covered in 70.5 minutes from the start. It was a 'Textbook' run, and such

was the expertise of the Enginemen that they made the job look easy. Rowe's tight stiff collar did not go limp with perspiration – only a little grubby in appearance. There was no doubt of his superb physical fitness as he followed the golden rule of firing 'little and often' but not always so little at that. A strong back and plenty of know-how made up a good top-link fireman following the motto 'pick the coal right, feed it right in the right place at the right time'.

Down the other side of Savernake we went like the wind, the 'King's' happy muted beat singing the song of her contentment. A mere feather of steam came from her bright copper-capped funnel and from her polished brass safety-valve casing as the quarter mile posts flicked past and overbridges shouted back. Country stations roused to our echoes as we clattered over the points and crossings. To me she was a live thing – a personality of all that had been built into her, with Braund and Rowe ministering to her needs that she might give of her best. Signals rushed past and let us by – Distant 'Off' Home 'Off' – mile after mile. Crossing Keepers and country signalmen waved the 'Limited' past. To be here on the footplate of such an engine was to feel part of a great railway tradition and an unequalled service.

We were really moving now. Woodborough – then Patney at 78 and on down the 1 in 122. 4.5 miles later we streaked through Lavington at 88 m.p.h. in 20% and the first valve on the regulator, gauge at 250. At this speed running was dead smooth, better than at 50 as we passed the up express from Plymouth due in London at 1.15 p.m. headed by King James II – 6011 on the buffer beam. A flash of green, some chocolate and cream streaks and she was gone. On our left above Edington and Bratton we could see the chalk white horse far up on the downs. Now it was time to shut off steam for Westbury. Brakes on and we swept the right-hand curve away from the bypass construction works. We came abreast of the illuminated '30' sign and swung into the left-hand curve through Westbury Station. Time 12.02.5 – 92.5 minutes for the 95.6 miles. We had covered the 15.3 miles from Woodborough to Milepost 94.25 in 11 minutes 15 seconds at an average speed of 83.3m.p.h. As we looked back the Weymouth coaches were slipped from the rear and we went forward with the load reduced to 445 tons.

Within 1.5 miles we were up to 60m.p.h. and skimmed 1,700 gallons of water from the Fairwood troughs before the slack to 30m.p.h. through Frome. By Woodlands Box we were all sat for the sharp 5 mile climb to MP 122.25 at Brewham finishing up at 1 in 107. 6024 barked her way up in 30% and only fell to 50 at the top – a superb climb. Now down the other side to Castle Cary at an easy 60m.p.h. Fred Sheldon tapped my arm and called "Look out, he's going to use the long one". Rowe took the 14 ft fire iron from the rack on the tender and with great dexterity turned it and thrust it into the firebox through the slit between the near-closed fire-doors and worked it between the firebars at the far end to remove clinker. He pulled it out white-hot and stacked it after turning without fouling anything either on board or outside.

Above:
6024 "KE1" with the down "Cornish Riviera Express" of 7th August 1933. The engine is in original condition. Note the footstep on the smokebox door: this was to enable firemen to reach the top lamp bracket. This was later moved on to the door itself. The train is passing Reading West. The passing time would be about 11.15 a.m. The livery is fully-lined G.W.R. Green. The tender bears the G.W. notation with full G.W.R. crest between the letters. The train is a perhaps unusually uniform rake of Collett 1928 coaching stock: it consists of probably 15 coaches including several slip coaches. The engine had recently undergone heavy intermediate repair. *[N.R.M. – M. Earley collection]*

We roared through Curry Rivell Junction at 72.5 and we were now on the Athelney Level in King Alfred's country. Through Keinton Mandeville at 69, Charlton Mackrell at 76.5 and up to Somerton tunnel which we entered at 75 to be surrounded by steam swept round the cab sides by the back draught, and lighted to flame colour by the fierce glow through the firehole doors. The noise was incredible to ears not used to it. The tunnel was over 1,000 yards long yet in a matter of seconds a pinpoint of light appeared ahead growing rapidly in size and suddenly we were out in the open again and all was quiet by comparison. Rowe closed the blower, checked the water level again and for a time duplicated Braund's observations of the signals. As each semaphore appeared he held his arm away from his body at an angle of 45 degrees imitating the position of the signal, and Braund nodded agreement.

The lines from Bristol came in on our right and the down track crossed overhead by a latticed skew-bridge – the 'Flyover' Junction at Cogload. Time was when it was a flat junction of some ill-fame as a cause of delays. At Creech Roe we skimmed another 1,400 gallons off the troughs at a steady 70m.p.h.

Next came Taunton – time 12.48.5. We were 3 minutes early. We were now all set for the 9.8 miles up to Whiteball. In the meantime our two coaches for Minehead were slipped and our load reduced further to 10 vehicles and 375 tons gross.

The four tracks from Cogload reduced to two as we swung past Norton Fitzwarren. Braund gave 6024 full regulator and proceeded gradually to adjust the valve gear from 20% to 22% as the gradient increased. 58m.p.h. through Wellington in 30% and on into the Blackdown Hills shouting her way up the 1 in 80 in 33%. I was thrilled by 6024's thrusting power output – 44m.p.h, 1990DHP. Braund gave his whole attention to the regulator in case she should start slipping on wet leaves, but she was as 'sure-footed' as ever. We entered Whiteball Tunnel as the gradient eased off to 1 in 125 and the speed recovered to 46 after a fine climb of some ten miles. And there was still 235 lbs on the gauge.

As soon as Braund shortened the cut-off and brought the regulator back to first valve, Rowe used the short iron to stir up the fire and break up clinker. Then he studied the injectors, damped down the coal and fired lightly all round the box and at the back. After the 40m.p.h. slack through Tiverton Junction, Cullompton was cleared at 75, by which time the pressure was up to the red line on the gauge.

Round a curve in open country, green, chocolate and cream with a plume of steam above it swept the up Torbay Express fast getting into its stride from Exeter with 13 on and whistling for Silverton – a grand sight – yet another King, this time Charles II from the number on the buffer beam.

Soon Braund made some light brake applications to steady her down as we approached a series of slight reverse curves that demanded respect. The rolling and lurching subsided as steam was all but shut off, speed falling gradually as we passed through Stoke Canon, heeling over on the super elevation of the rails and passing the confluence of the Culm and Exe rivers.

At Cowley Bridge Junction we passed the Exe again, and clattered over the points where the Southern tracks from Plymouth came in on our right. The next 'Home' Signal was off, but the Distant on as we drifted quietly along with brakes on to bring us to a smooth stop in Exeter St. Davids a full five minutes before time.

"And that's that" said Braund speaking for the first time since we left Paddington. The sudden silence was most striking. Our time was 164 minutes net for the 173. 3 miles – Time 1.18 p.m.

Braund and Fred Sheldon stepped down to have a look round the engine. Rowe put the blower on and added a few shovel-fulls to the fire, and then swept the footplate clean with a handbrush in real 'housewifely' fashion. He offered me a swig from the tea-bottle on the ledge above the firedoors, but I declined and said I was certain he was the more deserving of the two of us. He was at some pains to understand the interest I should have in travelling on the engine, and I pointed out that every boy worthy of the name carried this ambition in his heart, some until they grew up, and even after. He admitted that he too had always wanted to be an Engine Driver ever since he was a lad, but that the long apprenticeship required much patience.

Eventually Braund and Fred Sheldon climbed back on to the footplate following their inspection. They reported that everything was fine.

"That was a pretty fair trip" Sheldon commented. "We lose another coach here, but 9 is still a good load. And now you will see some really *Hard* Work".

The five minutes lead we had was used up in the stop and right on time at 1.26 the guard's whistle shrilled the "Right Away". And that was what Rowe shouted above the roar of the safety valves and Braund opened her up. There was a half-revolution slip of the driving wheels, checked by the regulator reduction to first valve. Then three-quarters open she really got hold of it and took us away on the level in 20%, quietening down happily on the first valve until we were bowling along at 60. Water was taken on for the last time at Exminster troughs, and as we swept along the Exe

Above:
At Bristol Temple Meads, 6024 is pictured having brought in the down "Bristolian" *[N.R.M. – Soole collection]*

Below left:
A fine photograph of 6024 passing Fairwood Junction nearby Westbury (Wilts) before the Second World War. She is hauling an up train of Collett coaches. *[N.R.M. – Soole Collection]*

river at Starcross the up 'Riviera' went by with a 'whoosh' nicely on time. After a short-lived acceleration to 75 Braund let her ease down to the restricted 60-62 required along the coast after Dawlish Warren.

Rowe's expertise and skill brought the fire and boiler to a state of readiness as we eased through Newton Abbot at 50, pounding out to Aller Junction where the Torbay lines diverged to the left. We swung right under a road bridge to face the 2.5 miles up to Dainton steepening to 1 in 42 and 1 in 36. A formidable proposition.

250 lbs on the gauge, water almost at the top of the glass and a thick clean fire burning furiously. We were all set for a virtually all-out attack.

Braund had given her full regulator, and now screwed her forward steadily to 50% cut-off – heavy working for a 'King'. But 6024 took it with complete authority and nothing flew out of the chimney except heavy exhaust. She gave us plenty to listen to as we roared up past Stoneycombe towards MP 216 with the echoes volleying back. Looking forward along the boiler at the rails curving upwards it was a fit setting for the drama being enacted by King Edward I – and all the course of a normal day's work. Speed at MP 216 fell to 46, and was down to 42 at MP 217, and finally to 36 as we entered Dainton Tunnel. Braund had been alert for any suggestion of slipping on that gruelling 1 in 36 climb, but as ever she was surefooted and took us over steadily. The water level fell dramatically on the 1 in 37 down the other side of Dainton , reason for the high level when we commenced the climb. 6024's last defiant blasting in the tunnel diminished as speed rose rapidly on the down grade on the 1st valve in 15%, 230 lbs on the gauge – the lowest for the trip, but just recall the tremendous effort involved.

Rowe now had five minutes to get his fire in order for the next great effort to Wrangaton. There was no clinker since he had used the long iron at Starcross, so he fired rapidly all round the box, and in the middle and the back as usual. We sped through Ashburton Junction at 60 then, all in order, Braund set her to attack the 9 mile climb to Rattery and Wrangaton in 45% full open. Rowe was ready. Full-bore up the 4.5 miles at 1 in 46 to Rattery where he kept filling any holes developing in the fire. Then King Edward was literally shouting to the high moors, and with the lowering cloud above and the wind coming in stiffly from the south-west it was an unforgettable experience. 6024 was a grand locomotive. Braund's old pipe gave off no smoke at all, but gave him something to clench his teeth on. Perhaps he liked the flavour. Curving sharply, first to right, then left the line swept upwards with the King's footplate going round at an ever shifting tangent. Here it was obvious why the G.W.R. kept to the 4-6 0 wheel arrangement. These sharp curves were no place for a 'Pacific'. From 48m.p.h. at MP 224 the speed fell slowly back to 40m.p.h at MP225. We cleared Rattery Rox at 37. Cut-off came back to 30% with the gauge at 235 as we barked our way through Marley tunnel accelerating to 42m.p.h. on the easier 1 in 120 grade. So, over the summit at Wrangaton at 50, gauge at 240 lbs. The worst was over. Rowe made his preparations for the end of the run, working the fire down to a clean red thin layer over the bars.

We were now crossing the southern spurs of Dartmoor. Braund controlled the train beautifully through the sharp reverse curves over the viaducts of Ivybridge and Slade, and up the short slope to the top of Hemerdon. Rowe took another drink and put on a 'smoke'. Speed was restrained to no more than 56m.p.h. down the 1 in 42. Then another climb for 1.75 miles to Mutley Tunnel which served to bring us down to 30m.p.h. Brakes on, regulator shut we rolled to a slow stop in Plymouth North Road 2.30p.m. Total net time 230 minutes. From Paddington a distance of 225.7 miles. Looking back on the tender I could see that Rowe had got to the slack at the bottom, having shifted 3.5 tons of coal to the firebox in the 210 miles to Brent where firing ceased. This was almost a ton an hour average for nearly four hours. A good morning's work by any standards. Water consumption totted up to 8,000 gallons and we had 1,900 gallons left in the tank. This equals 35 gallons per mile. It was a good run and a good finish. My respects to 6024, King Edward I and her crew".

The first heavy general overhaul of the locomotive was carried out at Swindon from 9th November 1934 to 24th January 1935.

General overhauls followed in 1938, 1944 and the last GWR heavy general overhaul in 1947 nearly at the end of the life of the Great Western Railway. By then 6024's mileage had risen to 864,202.

The general repairs of 1938 had included replacement of the right-hand outside cylinder. At a light repair at Swindon in 1945 the two inside cylinders were renewed at the reduced diameter of 16".

During all of this time the loco was confined to the GWR main lines to Bristol both via Bath and Badminton, the line via Westbury to Exeter and Plymouth and to Birmingham Snow Hill and Wolverhampton. It hauled the heaviest and fastest trains on the line. During this period 6024 was kept at Plymouth Laira Shed – July 1930 to January 1934; to December 1948 Newton Abbot; to August 1954 Laira; to September 1961 Old Oak Common; from September 1961 to June 1962, Cardiff Canton.

The engine records show that the locomotive was not immune from various removals from traffic during this period and from time to time it was stopped either for repairs at Newton Abbot or Swindon as necessity dictated.

What of the competition during the 1930s and 1940s?

In 1933 William Stanier left his post as Assistant Mechanical Engineer at Swindon to join the London Midland and Scottish Railway as Chief Mechanical Engineer at Crewe. He took over from Henry Fowler who had not long before brought about the introduction of the "Royal Scot" class of 4-6-0 locomotives. Mr Fowler had done this after trials with "Launceston Castle" from the GWR which was run on the main line between Euston and Carlisle and was said to have amply proved to the directors of the LMS that the policy of using numerous small engines rather than larger locomotives was simply outdated.

Stanier's designs signalled a permanent end to the "small locomotive policy" of the LMS. His largest locomitves were of course the "Princess" and "Duchess" Pacific class locomotives.

These designs brought much of the Swindon engineering practice to the LMS for the first time including for instance the building of locomotives with tapered boilers. He broke from the GWR, reluctance to build Pacific wheel arrangement locomotives probably stemming from difficulties experienced with "The Great Bear". The advantage of building the larger design of 4-6-2 wheel arrangement engines was partly, of course, to spread the weight over a larger number of axles without the inconvenience of suffering a longer fixed wheelbase brought about by the use of a locomotive with eight coupled wheels. The use of a rear bogie also enabled the firebox to be widened near the outside limit of the frames on each side rather than shoehorned between the frames. The "Kings" had been the most powerful narrow firebox locomotives built in this country but of necessity the width of the frames constricted the firebox area available for heating the water. The fireman does not need such a heavy throw of coal to reach the front of the firebox in a wider and shorter box either.

It is also noticeable that Stanier used outside valve gear (four sets were used in his "Princess" Pacific class locomotives) instead of two in the GWR four-cylinder locomotives.

Stanier did however, keep to the four-cylinder arrangement for his largest engines but also designed slightly smaller locomotives with three cylinders instead of four. The GWR never built a three-cylinder locomotive.

Meanwhile, the LNER "Pacific" locomotives designed by Sir Nigel Gresley were built with two sets of outside valve gear but no inside valve gear. the inside valves on the three cylinder locomotives were driven by Gresley's own design of derived motion on a similar principle to the GWR rocking shafts.

Meanwhile, by 1939 the GWR was beginning to appreciate certain shortcomings in the design of the "Kings". A report dated 19th May 1939 from Swindon Drawing Office referred to some of these points and drew conclusions from work on compound locomotives in France. In its preface entitled "Outline of Recent Development", three important modifications were mentioned; stating "... The effect has been to increase the indicated horsepower per ton weight from 23 to 37... This 61% increase has been achieved by the following comparatively slight modifications:

Fitting "Kylchap" double blast pipe.
Increasing the degree of superheat.
Increasing the size of all steam passages..."

The implications for GWR practice were being clearly underlined. Presumably further work was stopped once the Second World War started three and a half months later.

The writer has not seen any wartime photographs of 6024. Photography of railway equipment was forbidden in wartime as a security risk. The locomotive was painted in unlined green, as an economy measure. The cab window was plated over with a blank sheet of steel to cut out the glare of the fire to outside and thus its visibility to enemy planes.

The railways were under considerable pressure in all departments in the War and there seems no doubt that maintenance standards suffered whilst the engines had huge demands made on them both to haul fewer but heavier trains than was the norm before the War.

Above:
Shortly before World War 2 started, "KE1" 6024 is firstly depicted at Platform 2, Paddington Station in outstanding condition: even the bright steelwork is immaculate. Note the clerestory coach in the right background behind 6024's tender.
[C. Bertram Mullinger]

5: Postwar developments

On the 1st January 1948 British Railways took over the assets of the GWR and this included 6024.

In the first year of nationalisation locomotive exchange trials were held. A number of other writers have much more fully described the exchanges including:

Cecil J. Allen, in his book on the subject; O.S Nock in his history of British Locomotives of the 20th Century. More recently "Steam Railway" and Adrian Vaughan have given more detailed descriptions of what happened.

The GWR sent 6018 "King Henry VI" to the ER where it ran from Kings Cross to Leeds on a regular basis during 1948.

The reason for the trials being held seems to have been firstly that now that one authority was running the railways it was sensible to standardise design so as to simplify maintenance, cut costs and ultimately also to provide locomotives common to all regions.

It was also recognised that the quality of coal available had dropped during and after the Second World War. Britain was exporting an appreciable amount of its production and indeed in the 1950s the Western Region ran its locomotives on quite appreciable amounts of Polish and French imported coals.

No 6018 did not shine in the coal consumption tables when comparison was made of the performances of the various companies passenger locomotives, but in fairness it was the oldest design of its group.

Later design had placed particular emphasis upon free steaming by attention to the input and exhaust of steam from the cylinders. It was also pointed out in defence of the engine that it was not run on the Welsh coal for which it was designed but harder Yorkshire coal.

Following these trials, B.R. continued to build some further examples of the "Castle" class at Swindon which were ordered before the B.R. designs were introduced from 1951.

Visually and for practical purposes one of the greatest contrasts was the accessibility of many more parts on the B.R. standard designs. Many thought some of the designs unattractive because they seemed to adhere more to American practice with much outside pipework and high running plates. The locomotives appeared very different to the designs produced by the GWR where many working parts and pipework were concealed under or between the frames and beneath sections of boiler cladding.

More emphasis was placed on a higher degree of superheating in the boilers of the B.R. designs although the last "Castles" built at Swindon were built with the modifications later incorporated in many B.R. locos including mechanical lubricators and superheaters of 32 elements. Double

Below:
6024 "KE1" is shown in Sonning Cutting on 19.4.54. This is the second photo in the book of 6024 running with the improved superheated boiler, 6 months after it was fitted during heavy general overhaul in August/September 1953.
[R. Blenkinsop]

chimneys were fitted to "Kings" and "Castles" from 1956.

Nevertheless, B.R. days did not signal the end of efforts to improve the performance of the "Kings". As long as steam power was being produced in the 1950s and after the modernisation plan of 1955, the order of the day seemed to be some continued research into improving the efficiency of some engines.

In the "Kings" the first significant modification was the replacement of the existing "King" boilers in 1952-1955 following trials with No 6022 in 1947. The number of small tubes was reduced from 171 to 115 to accommodate the additional row of superheater flues to produce a 4-row superheater, thus providing the higher degree of superheat possible with improved lubricating oils.

The new boilers were the same size externally as the existing No 12s. The barrel was 16' long and the firebox casing was on the outside 11'6" long. These new boilers gave an immediate benefit in fuel economy.

The works records show that No 6024 left Swindon on the 11th September 1953 with the first of such boilers she was to carry. Four in all were fitted at different times. The last, boiler No 8610, was fitted in June 1960. The boiler had been built in February 1953. From June 1958 until June 1960 it had been carried by No 6000.

In early B.R. days, different liveries were applied. Some of the large express passenger locomotives were treated to a light blue livery. In his book "Portrait of Kings" Kenneth Leech records that when he saw 6024 at Bath on the 18th November 1950, the locomotive carried blue livery. That was shortlived and the later fully lined out Brunswick Green B.R. livery was applied within two years.

Various designs of blast pipe and double chimney were tried and Kenneth Leech is one of the authors who describes the various elements in more detail. Swindon designers, led by R.A. Smeddle, settled on a cast-iron double chimney. 6024 acquired a double chimney in March 1957.

Mr Leech's record of 6024 contains a photograph taken on the 22nd March 1957 and shows the double chimney at that time.

The new chimneys helped substantially to reduce one of the long known vices of the class, which together with improved double blast pipe design helped the locomotive to expel steam and exhaust more efficiently. The vice in question was one which impeded the efficiency of the locomotive to an increasing extent as speed rose above 20 m.p.h.: back pressure in the cylinders. The cylinders were, because of this, unable to run freely because the engine's capacity to expel used steam was not being exploited fully, but instead inhibited by the restricted steam passages to the atmosphere.

The increased blast area also

Above:
6024 is seen after repair at Swindon Works on a turntable road. The neighbouring pannier tank No 4692 appears to have had a major overhaul and its paintwork contrasts unfavourably with 6024. Although carrying 83D Plymouth Laira shedcode, No 6024 has been withdrawn from traffic at Old Oak Common shed in June. *[K. Leech]*

Above right:
West of Wellington 6024 is climbing towards Whiteball Tunnel with the 3.30 p.m. Paddington to Plymouth express, on the 23rd August 1954. *[R. Blenkinsop]*

Right:
31st August 1954. and Dick Blenkinsop was in the Midlands. 6024 is climbing Hatton Bank with the 9.10 a.m. Paddington to Birkenhead express. The loco appears to have a good reserve of coal. The photograph is so clear that it is possible to count most of the larger lumps! This is a scene which has recently been reproduced almost exactly! *[R. Blenkinsop]*

6024

6024
KING EDWARD I

THE BRISTOLIAN
116

Above:

Having made a fine climb of Hatton bank, No 6024 races past with the Didcot to Derby leg of the main line charter working on 12 January 1991. *[Peter Nash]*

Above left:

6024 was out of traffic for heavy general repair from 16.1.57 to 20.3.57. Whilst at Swindon the locomotive received a double chimney, the lubricator was moved to its present position forward of the outside right-hand steam pipe and a different boiler, No C8605, was fitted. The picture depicts the engine at Chippenham on a running-in train on 22nd March 1957. The engine is in ex-works condition and the paintwork is almost spotless. Some of the alterations are visible, the "SC" denoting the newly-fitted, self-cleaning smokebox, beneath the shedplate, 81A Old Oak Common having been the engine's home shed since 1954. *[K. Leech]*

Left:

On 11th June 1957 shortly after overhaul, 6024 is in fine condition as she runs at speed on the approach to Twyford station with the down "Bristolian". *[R. Blenkinsop]*

improved steam production by the boiler.

In the mid 1950s it was found that the age of the "Kings" was telling in wear on the front ends of the locomotives and as heavy repairs came up, the cylinders were removed and the front sections of framing cut out and renewed. It seems possible that 6024 underwent surgery in this way in the first three months of 1957 during a heavy general repair at Swindon.

It was probably because of the "Kings" status as the Western Region "flagships" that Swindon took an especial interest in them and was especially sensitive about their performance. The trains they hauled too were those most likely to receive public attention, so timetable schedules had to be kept.

In the 1930's the public had been very interested in the increasing speeds by which man could travel by both land and air. Various air races were held. On land, motor cars were run up to over 200 mph at such places as Pendine Sands in South Wales and in the United States. It was obviously good publicity for the railways to produce their own speed records and the GWR did not keep out of the limelight in this way, although it never sought to beat the other speed records established by the LMS and LNER, concentrating more on high average speeds as with the performances of the "Cheltenham Flyer".

In the 1950's, economy ceased to be the overriding watchword as the end of rationing came about and the country moved out of the shadow of the post-war years so that the performances of the locomotives could again receive some attention. The wartime 60mph limit was also lifted.

Beyond the other modifications received by the engines, mechanical lubricators were installed in place of the original front end cylinder and valve lubrication fed from the cab by sight feed displacement lubricator. This was to ensure that not only was there sufficient front end lubrication, but not so as to be excessive leading to carbonisation of the ports. The engines were vulnerable to this with the advent of higher superheat, if over-oiled.

Initially, the lubricator was placed behind the outside elbow steam pipes and later they were moved to a position on the right hand foot plate in front of these pipes. Steam pipes were found to occasionally fracture and these were modified during the blast pipe modifications to produce a smoother flow, with a more pronounced "horizontal" mid-point of the pipe brought about by their twin larger radii.

Efforts were made to install a self-cleaning smokebox. Subsequently the original wire screens were changed to "a wire basket" completely enclosing the blast pipe which operated as a spark arrester.

In January and February 1956, following an accident with No 60700 on the LNER, some of the "Kings" bogies were found to be fractured and the whole class was withdrawn in January and February 1956. During this period, the WR borrowed four "Pacific" class

locomotives from the LMR to cover the withdrawal of the "Kings" whilst the "Kings" bogies had strengthening pieces welded to them.

The cabs were fitted with ventilators in the roofs. Those ventilators had firstly only been installed on the one streamline "King" No 6014. A square section step under the smokebox door was installed in place of the previous rounded sheet of metal over the inside cylinder valves. This was to prevent staff slipping when attending to the smokebox. In the later 1950s the work of Messrs Smeddle and Ell at the locomotive department at Swindon gave rise to plans to alter and streamline the steam passages to and from the cylinders of the "Kings". It is said that they would have improved economy and speed from such work. Additionally, some roller bearing sets were delivered but never fitted. Unfortunately, nothing further was done because main line diesels were introduced in quantity from the late 1950s and the "Kings" were gradually displaced from their top-class work.

Nevertheless, during the late 1950s the "Kings" were usually employed on the heaviest and most important Western Region trains to the South West and Bristol. By now, too, enthusi-

Above left:
A view of 6024 "KE1" during its last years on West of England expresses; on 18.7.58 6024 is seen hauling the 7.30 a.m. Saturdays only Paddington to Kingswear express along a favourite place to watch trains: the sea wall at Teignmouth. *[R.C. Riley]*

Left:
On 6th August 1960, while 6024 was being used regularly on trains to the North West, she is leaving Paddington with the 11.10 a.m. Paddington to Wolverhampton (low level) train and passing under Westbourne Bridge. This was one of Dick Riley's regular vantage points for his photography. It provided an excellent view of the station approaches, particularly departing expresses accelerating away towards Old Oak Common. *[R.C. Riley]*

Above:
6024 is depicted at Seer Green in Buckinghamshire on the down 3.30 p.m. Paddington to Wolverhampton express. Between here and West Ruislip the line runs gently downhill in the London direction. Engines often reached high speeds hereabouts and "KE1" herself attained the speed of 102 mph(see text). *[R. Blenkinsop]*

Right:
23.5.61. A further shot of 6024 on an up train (rather than a down run referred to in this article) approaching Leamington Spa locomotive shed. A "Prairie" tank and "Hall" 4-6-0 loco stand outside. Still with a year's ordinary service to go, 6024 looks very well cared for. *[R. Blenkinsop]*

asts had the bonus of chocolate and cream coloured coaches on some express passenger trains behind "Kings".

By 1962 the "Kings" were completely supplanted by "Warship" and "Western" diesel locomotives on lines on which they had run for 35 years.

It was at this time that Swindon works abandoned further improvements which had been considered for the "Kings" including plans to fit roller-bearing axleboxes. Nevertheless, 6024 ran over 100 m.p.h. on several recorded occasions in B.R. days.

Whilst running on the "Bristolian" in 1957 Baron Vuillet recorded a top speed of 101.5 m.p.h. with No 6024 between Wootton Bassett and Chippenham on Dauntsey Bank.

Probably between 1960 and mid 1962 on what must have been one of the engine's swansongs on the "Cambrian Coast Express", "King Edward I" reached 102 m.p.h. on the fast stretch of line near Denham in Buckinghamshire. The engine was hauling 7 coaches of 234 tons tare, 250 tons gross. Additionally the engine is recorded as passing Bicester at 91 m.p.h., Blackthorn at 88 m.p.h., Princes Risborough at 71 m.p.h. and past Saunderton Summit at 61 m.p.h. This run is recorded on page 316 in "Modern Railways" of November 1962.

No 6024, in common with others of its class was transferred from the West of England line to West Midlands trains. It hauled the "Cambrian Coast Express" (as far as Birmingham and Wolverhampton) and the "Inter City" regularly.

We are fortunate to have a run recorded on the 19th November 1959 by R.C. Riley, on the footplate of 6024.

The train was the 8.30 a.m. from Paddington to Wolverhampton; 6024 hauled a load of 9 coaches, weighing 325 tons gross. The engine was noted to be in less than perfect condition, having run 150,000 miles since its last heavy overhaul. The performance was rather in keeping with the engine's condition. The train reached Wolverhampton Low Station 5 minutes late. The highest speed of all was attained between Fenny Compton and Leamington Spa.

The full record of the run as it originally appeared in "Trains Illustrated" for February 1960 is reprinted fully (by kind permission of R.C. Riley and Ian Allan Ltd)and starts on page 26.

No 6024 "King Edward I" back on the main line and at work in 1990/91

Above:
Climbing Wilmcote bank out of Stratford-upon-Avon with the 17.05 Stratford–Tyseley of 15 April 1990.

Left:
Nearing Moreton-on-Lugg when working Hereford–Shrewsbury on 24 June 1990.

Above right:
Storming past Cropredy, north of Banbury with the Didcot-Derby section of the 24 November 1990 train.

Right:
About to pass under the A41 road bridge north of Aynho Junction with a BR private charter train on 19 May 1990.

[All: Peter Nash]

THE ROYAL VENTURER
Z57
6024

6024

"In order to obtain first-hand experience of the operation of the new services, I made a footplate journey on the only down two hour train to survive in the current timetable, the 8.30 a.m. Paddington, a new train put on to offset the withdrawal of the 8.40 a.m. Euston. It is also the first non-stop train to run between Paddington and Birmingham since before the war, the other recent two hour trains, now decelerated, the 9.0 a.m. and 10.10 a.m. Paddington having made brief stops at High Wycombe and Banbury respectively. On the face of it, therefore, a "King" on nine coaches appears to have an easy task, and it is allowed 0.5 minute longer to pass Ashendon Junction than the old 9.0 a.m. with its one minute stop at High Wycombe. The engine provided was 6024 King Edward I in charge of Driver Bailes and Fireman Jeffery of Old Oak Common, and I was accompanied by Inspector Hancock. Although it was only the second Monday on which the train had run, it was well patronised, there being 60 First and 165 Second class passengers, bringing the loaded weight to about 325 tons. The start was quiet and restrained, This is a feature common to all direct Birmingham line trains out of Paddington, for they must cross from main to relief line at Subway Junction, Ladbroke Grove or Old Oak Common West, and there is a 40 m.p.h. speed restriction at the latter place in any case, so engines are not usually opened up until North Acton is passed. Slow running in the early stages of the journey is often the cause of a minute or two lost, but this can be made up at later stages of the journey. My run was no exception, and it was soon evident by the amount of steam leaking at the front end that 6024 was not in the best of condition. I later learnt that it had done 58,000 miles since an intermediate repair a year previously. As it must therefore have covered about 150,000 miles since its last general overhaul, its performance was a very creditable one.

A 40 m.p.h. slack at West Ruislip, quickly followed by a 30 m.p.h slack at Gerrards Cross increased our lateness to four minutes at High Wycombe. This far the line was built to main line standards, but thence to Princes Risborough it covers a former branch brought up to main line standards, and the difference soon became evident. Through the curves of High Wycombe there is a speed restriction of 35 m.p.h and at West Wycombe a similar restriction of 45 m.p.h. with seven miles of adverse gradient to follow, such slowings do not help matters, and when the summit is reached beyond Saunderton there is no question of fast running for the old single line curves round very sharply to the Princes Risborough approach with a speed restriction of 60 m.p.h. When the line was doubled in 1910, the new up line for 2.5 miles between Princes Risborough and Saunderton was built apart from the down line, so enabling the gradient and curves to be eased. I would not describe 6024 as rough riding, but it was certainly a hard riding engine and the wisdom of Inspector Hancock's advice to "hold tight" over this stretch soon became apparent. Hampered as we were in the early stages of the journey our maximum speed had been a brief 64 on the downhill run from Beaconsfield, but now 6024 had a chance to run as the curves past Ashendon Junction have recently been eased and the one time 50 m.p.h. speed restriction lifted. Through the closed station at Blackthorn 6024 was up to 75 m.p.h. in readiness for the climb to Ardley, and was still maintaining 64 through Bicester North, the only station of any consequence on the 1910 "cut-off" line. Although still two minutes down the Ardley signalman was not ready for us and we were brought to a brief stop outside the station on the 1 in 200 adverse grade. Thence to Aynho Junction we took things easily for the flyover junction by which the down line joins the old route via Oxford is restricted to 60 m.p.h. Through the fine new station at Banbury down trains are restricted to 75, but further signal checks south of the station considerably reduced our speed, although we had to slow down in any case for a 15 m.p.h. relaying slack at Banbury Junction. After these setbacks 6024 hammered away at the climb through Cropredy, passed at 64, and speed steadily increased on the down grade through Fenny Compton and Southam Road and Harbury to reach 82 before brakes were applied in readiness for the 40 m.p.h. restriction through the middle road at Leamington, passed nine minutes late. 6024 was now opened up for the climb to Hatton, with 25% cut off at first, eased to 20% passing Warwick. The "King" responded well to this treatment and would have maintained 60 m.p.h. up the steepest part had it not been for a brief bout of slipping, quickly corrected by Driver Bailes ("Autumn Leaves," he remarked laconically). We made a fast approach through the Birmingham suburbs, reaching 78 at Olton, and thanks to two minutes regained by the engine and three minutes recovery time allowed, we pulled up in Snow Hill only four minutes late...."

Above:
10th July 1960: "KE1" passes Southall on the down midday milk empties train. The photograph depicts other features of historical interest! To the right of the rear of 6024's train is one of the G.W.R. AEC diesel railcars. The shed and yard at Southall have been sold and the destination of the B.R. single railcar is "Staines West". The Staines West branch is closed although part is used for aviation fuel deliveries to Heathrow Airport and runs through the M4/M25 interchange. Overhead electric lines are due to take trains from Paddington to Heathrow along here in the early l990s. *[M. Pope]*

Left:
6024 is depicted at Shrewsbury in post-1957 period. Judging by the level of coal in the tender and the torrent of steam from the safety valves, she is about to run to the station whence she will take the "Cambrian Coast Express" on to Paddington. [G.W. Sharpe]

Below left:
"KE1" had by now been transferred from Plymouth to London; she is pictured at Old Oak Common shed in February 1959. Later in the year the last heavy overhaul of the engine began. Behind 6024 is one of the original series of "Hall 4-6-0"'s. In 1954 there were no less than 17 "Kings" based at Old Oak Common.
[G.W. Sharpe]

The record originally appeared as part of a longer article describing the effects of the then diversion of London Birmingham passenger traffic from the LMS Euston line to the W.R. Paddington to Snow Hill, Birmingham route.

These trains were the swansongs of the "Kings".

The "Kings" were used for a short period during the late 1950s and early 1960s on express trains between Cardiff and Paddington. 6024 was transferred to Canton Shed in September 1961. It had been used on trains to South Wales before that time! "Trains Illustrated" for February 1960 records the engine's work in December 1959 on trains from Cardiff to London thus: ".... From November 25 the balancing working of the 12.45 a.m. Paddington – Cardiff newspapers was altered from the 8.15 a.m. to the 8.00 a.m. back from Cardiff, the "Capitals United Express" and the diagram was handed over to Old Oak "Kings" ... The first choice on November 25 was a "King" due for the shops, No 6024 which arrived in Cardiff 25 minutes late with the newspaper train and steamed so badly on the 8.00 a.m. up as to reach Paddington 43 minutes late ...." On 30th November 7029 "Clun Castle" took this train into Paddington on time but ... the nadir was reached with a late arrival of 60 minutes in London by No 6024 with the 8.00 a.m. on December 7 ...."

The engine was close to a heavy overhaul,and the quality of its work probably reflected this. This last overhaul at Swindon was carried out in 1960.

In the summer of 1961 Colin (L.C.) Jacks, then a passed fireman rode on 6024 between Lapworth and Birmingham Snow Hill. Here is his description of the run.

"My only trip 'in charge' of a King was in the summer of 1961 when as a Passed Fireman, that is a fireman passed by the Inspector at Swindon on the rules and regulations, having a sound knowledge of all the working parts of the steam engine, thus being able to be upgraded to do driving work.

I was given the job to 'pilot' the driver of the down 'InterCity' the 9.00 a.m. ex-Paddington over the down relief line from Lapworth to Snow Hill, the Old Oak driver not signing the road down the relief line. The down main line, the usual route, was occupied by a track ballasting machine.

Arriving at Lapworth in good time I presented myself to the signalman and told him what I was there for. He was disgusted that I should have to 'hold the man's hand down the relief'. He was also annoyed at the fact that he had got to stop such an important train at his signalbox, and added that any track work should be done at weekends. He further indicated that I did not look a suitable man to do such an important task, (I was 28 years of age).

With this sort of atmosphere prevailing I told the man that I would be on the down platform till such times as my train was due and left his signalbox.

I made myself comfortable on a platform bench, and enjoyed the peaceful rural surroundings, broken only by the passing of trains.

Just when I thought my train should be showing up I heard an engine's whistle being blown protestingly at the danger signal, this signal was lowered, the signalman came to the window and pointed towards the train, and slowly down the platform rolled the Inter City.

6024 King Edward I was stopped with inch-like precision opposite me. I must admit I had a few butterflies here and there. What with me looking up at this huge engine and the 'cockney' fireman looking down at me I felt a trifle pathetic.

"Up you come me old mate" he said, I climbed aboard into that most sought after place, the drivers side of a King's footplate. "Ten on matey" the driver said,

"She's OK, look after her", and deposited himself on the firemans seat.

Never did the driving side look so big and empty, I bravely went over and promptly tripped up the raised footplating where the driver normally stands. I regained my poise, opened the large ejector, checked that the down main to down relief signal was off, blew the whistle, opened the regulator and 6024 moved majestically off.

As we went by the box the signalman shouted to us that we were right away. Once on to the down relief line I opened the second valve of the regulator, pulled the lever back and this took us along in fine style. The fireman didn't seem overworked so I considered I was driving my first King okay.

Through Knowle and Dorridge and on past Bentley Heath we started to get up to the lines maximum speed of 60 mph, the next distant signal, that of Solihull I expected to find at caution, but when we ran over the ramp the ATC bell rang loud and clear, so much for issuing of special notices that said we would rejoin the mainline at this point. Still we pressed on and from this point the lines start to fall gently. I put the regulator on the first valve and the safety valves started to blow off which prompted the fireman to drop the firehole flap and the driver to remark that we really didn't need all that steam at the moment, in a slightly sarcastic manner. The fireman's amusing reply to him has no need to be quoted here.

Past Olton and through Acocks Green then right away Tyseley, running through here I hung my head over the side. It's funny there's never anyone about on the platform when you want to be seen. No doubt we would carry on down the relief line and go over on to the mainline at Moor Street.

Approaching Small Heath I shut the regulator and put the lever in the 45% cut off position. 6024 bucked and bronked her way over the crossing as if protesting about being on the relief line.

Passing Bordesley I judged that the speed of our train would drop down to the required 25 mph to cross from the relief to the mainline at Moor Street, the hump on which Bordesley station stands on did this trick quite nicely.

As we approached the crossover, I opened the regulator, dropped the lever into the 60% cut off, the Old Oak driver came across to me and asked me if we would get up the tunnel at this speed, Snow Hill tunnel being some 400 yards long with a rising gradient of 1 in 45, a steep bank indeed. "We'll be okay" I bravely said. 6024 responded magnifi-

Top:
6024 dumped in a siding at Cardiff Canton Shed in the week of her withdrawal. With 6024 is No 6003 "King George IV", withdrawn at the same time after running 1.92 million miles. 2nd June 1962, "KEI"'s bunker is full of coal the chimney remains covered; the engine was not to steam again until February 1989.
[G. Morrison]

Above:
On the 23rd September 1962 "KE1" is shown at Swindon yard tenderless, behind 6023. All visible parts are in place including connecting rods which were cut through before the engines reached Barry.

Left:
Contrasting with the dereliction of 1962, No 6024 looks superb when restored and seen at Quainton Road on 24 September 1989. *[Colin Washbourne]*

cently when I opened the second valve of the regulator as we plunged into the dark smokey bore and dug our heels into the steep bank. The noise was tremendous, 6024 kept her feet well, but then all Great Western engines were sure footed. Soon we were blasting our way over the ATC ramp that was virtually at the top of the rise in the tunnel and as we got on to the level I put her on the first valve of the regulator and we ran down No 5 platform sedately as befits a King class locomotive and train.

We came to a stand just a few minutes late, I said my farewell to the Old Oak men and got off 6024's footplate feeling somewhat pleased with myself on the way I handled my first King hauled express.

The Paddington guard asked me if I had seen the pilotman of 'his' train. When I told him that I was one and the same person he looked somewhat displeased that the likes of me should have driven his train, I felt a dent in my ego, "What's the name then?" he snapped – "Driver Jacks' Tyseley Loco", I proudly said!"

The engine's use in this period was not confined to the busiest and heaviest express trains. A photograph in 1960 shows the engine at Southall on a milk empties train.

In February 1962 No 6006 was the first "King" to be condemned and was cut up.

Nos 6023 and 6024 were at Canton Shed in June 1962 when they were condemned and by September 1962 13 more "Kings" had been condemned. Nos 6005 and 6028 were condemned at Old Oak Common in November 1962 and the last four followed in December 1962 – 6000, 6011, 6018 and 6025.

In February 1963 "The Railway Magazine" referred to the haulage by 6018 of an enthusiast special train from Paddington to Wolverhampton on the 17th November 1962. It went on to say that on the 9th December 1962, 11 of the class were seen at Old Oak Common Shed. One of the magazine's correspondents saw three of the class at Paddington on the 27th October 1962. Nos 6000, 6005 and 6011 had all arrived on Newbury race specials and the report concluded "... it is doubtful if such a sight will ever be available again". He may have despaired prematurely!

The final run by a "King" prior to the return of 6000 in 1971, was by 6018 on an enthusiasts' special from Birmingham to Swindon on the 28th April 1963.

From Cardiff, Nos 6023 and 6024 were not sent directly to a scrapyard as had been some other engines, but to Swindon works. There they were kept separate from their tenders but not scrapped and on the 10th October 1962 British Railways sold them to Wards Scrap Dealers of Briton Ferry near Neath in South Wales with their tenders. The engines were removed to Cardiff with their tenders but were barred by B.R. from travelling through Glamorgan to Neath because they were too heavy for the line west of Cardiff. A month later on the 26th November 1962 the engines were resold to Woodham Brothers, scrapyard at Barry and were delivered there but not cut up. It is well known that this chance resale saved the pair for posterity because of course Woodham's hardly cut up any of the engines they received into their yard. At some time after arriving at Swindon works though, Nos 6024 and 6023 suffered the burning through of the piston, eccentric and connecting rods also lower slide bars which made for easier towing of the engine whilst 'cold'. This was prior to their despatch to South Wales.

More of this later.

Not only did Mr Woodham appear to have nurtured the wishes of others to rescue locomotives from his yard, but has also said it was more economical for him to scrap British Railways' wagons whilst he could buy them and leave the engines intact. It is now well known that all but about 20 of the engines he had bought from B.R. have been 'rescued', removed from his yard and taken to other places in the country for restoration – a few have been dismantled for spares.

In many cases the engines are, or have been, in a condition which was a considerable challenge to the resourcefulness and resources of the buyers. Not least among these was "King Edward I" on purchase from Barry in 1973.

6: The end of steam and start of preservation

One of the recent books telling the stories of Barry engines points out that a number of classes of British steam locomotives would now be extinct were it not for the Woodham Brothers' yard. There would be no "Halls", "72xx" 2-8-2T, Southern "Q" 0-6-0 or Somerset & Dorset 2-8-0.

Although 6024 does not fall into this catagory, the wider question raised by the 200-odd Barry rescues is why, if none at all of many classes was originally preserved in 1962 – 1968, why are Societies and individuals prepared to go even further than secure one example of each class?

In 1963 there seemed a real chance that No 6018 of the "King" class would also survive. It appears Butlins wished to purchase it and an overhaul began at Swindon works but eventually another locomotive was purchased instead. It is alleged that British Railways asked too high a price for the engine and after overhaul had commenced, work ceased and it was cut up.

Many enthusiasts bemoan the absence of "47" and "Grange" classes from GWR stock.

From the 1989 viewpoint it is still extremely surprising that there was no other member of the "Kings" preserved and the same can be said for the "LNER A3"'s. If the opportunity were now to present itself, it seems likely that possibly another "King" in complete running order would be bought somehow. The cost in 1963 would have probably been between £3,000 and £4,000 which was the price of a good house then. That would probably mean £100,000 to £150,000 in 1989.

Whereas for some reason enthusiasts of 1963 couldn't or wouldn't buy engines, those of a generation later have bought and restored engines with 20 or 25 years accumulated corrosion, neglect and loss of parts both steel and nonferrous.

In 1963 it would certainly have been known that B.R. intended to preserve a number of notable engines. They had started doing so several years before. "Lode Star" at Swindon is one example. Their list was very selective though and even altered at times – it seems that "Duke of Gloucester" and "Ben Alder" in Scotland were to be preserved, but were scrapped, although the former has survived.

Many omissions were viewed with dismay by many people. No "A3"'s were included in the B.R. list and Alan Pegler purchased 4472.

By and large though, most enthusiasts seemed unwilling or unable to organise preservation schemes or come up with the necessary money.

In "The Railway Magazine" of February 1963 a letter was published from Mr R. Bell the secretary of the "Princess Elizabeth" Preservation Fund pointing out that "as a second object we hope to preserve 'The Royal Scot' which we were offered for £1,900. We are extremely sorry to report that there has been quite insufficient financial support ..."

Nevertheless, the same issue carried news of reasonable progress in raising money for the purchase of No 61572, the last "B12" 4-6-0.

With hindsight and the benefit of our experience since 1968 it is easy to see some of the possible reasons for the somewhat dismal state of affairs in 1963. First and foremost the end of British steam was still 5 years away. There were still a lot of steam locomotives to be seen at London termini.

Above:
On 18th October 1964 "KE1" is in a siding at Woodham Brothers scrapyard at Barry. She is stored behind 6023 and ominously close to a wagonload of scrap. Gauges and the driver's brake valve have been removed, so has the safety valve cover. Later the chimney was to be transferred to 6000 "King George V". *[Mr Down]*

Above right:
This photograph was taken in 1972. By this time Chris Tankard and fellow founder members had formed what was then the "King Preservation Society" and reserved the engine for preservation. This left-hand side view shows that on this, the lower slide bar was missing. *[R. Thomas]*

Right:
This photograph shows 6024 at Barry again in 1972. There are many parts to replace down to cab window and spectacle plate frames and glass; on this right-hand side view the cladding sheet immediately in front of the cab has been removed to reveal copper pipework in situ beneath it. *[R. Thomas]*

Some schemes which banished steam engines from large areas were only in the early stage of execution or planning. The Euston and Bournemouth electrifications are examples.

Railway enthusiasts were probably of the opinion that their first priority was to continue watching engines rather than buying them.

In 1963 too, we lacked the examples of those who set the pace in the preservation world and showed what could be done with rusty equipment and derelict track after 1968. The closure of some of the most attractive lines had only just begun and the full impact of the governmental shrinking and transformation of the railways had probably not sunk in.

Preservation has, too, become an expensive affair. When 6024 has been completed the overall costs will have been more than £190,000 and at today's money values even more. In the early 1960's individuals had generally less money to spend on leisure pursuits and hobbies and working weeks were longer giving less time for amateur restoration activities. Moreover, we had not then before us, as now, the examples of many successful fund raising schemes where a few individuals may have begun a scheme with a "wing and a prayer" as ours did. We can now look back on successful share issues in locomotives such as "Port Line", "The Severn Valley Railway PLC" and outside railwayana "Band Aid" and other fund raising concerts in other fields of modern life.

The attitude was probably somewhat different 25 years ago in that very large preservation schemes were seen as ideal for big museums but beyond nearly all individuals.

The disappearance of steam traction had a profound effect on the emotions and memories of a large number of people too. There were no British railway steam runs on the standard gauge for over 3 years until the milestone of 6000's return in 1971.

Many thought that what had been commonplace had gone forever. Imagine the reaction of the country's anglers if all the fish disappeared overnight!

Lastly it is probably worth comparing the attitude of the public in 1963 to borrowing money and those of the banks in their lending habits. 25 years ago most people had no bank accounts and lending was far less widespread than now. The number of finance companies and smaller banks has proliferated since then. That trend began in the 1970s since when we have seen an explosion in the use of credit.

7: At Barry scrapyard

Whilst at Barry Nos 6023 and 6024 suffered the removal of many non-ferrous parts including safety valve bonnet, whistle, cab fittings and lubrication parts together with brass beading. Nevertheless, Michael Wyatt of the Great Western Society was one of those people who saw 6024 several times at Barry. He has said that he saw the engine there for the first time in April 1964. He said he obtained permission in writing from Mr Woodham to go and look at the engine on one Saturday afternoon. Both the "Kings" were at the scrapyard and both were very complete. He bought from Mr Woodham a safety valve bonnet from one of the engines but hadn't the means to bring it back to his home. He remembers that the engines carried their cab fittings as did many of the other engines at the yard. Some other engines still had cabside number plates.

The chimneys were still on the engines together with brass beadings and the front bogie axlebox covers which were later removed.

The lack of rods and other items removed meant that, to most, it was axiomatic that it would be beyond a preservation society to restore either locomotive.

In the case of 6023 a derailment in the yard produced a cutting through of the rear pair of driving wheels.

At the same time the "Kings" featured strongly in the memories of many railway enthusiasts and where there was a will there was a way. Sceptics it seems shook their heads when in 1972 Mr Chris Tankard formed the "King Preservation Society". By September of 1972 the society had 60 members. No 6024 was the locomotive which the society wanted to buy. The task was considered a "mission impossible" with no connecting rods.

With Woodhams' consent, parts were removed from No 6023 including the chimney (6024's is on No 6000) and other preparations for its removal took place.

Chris Tankard provided particular financial impetus by way of a loan and eventually No 6024 was bought by the society for £4,250 which included the cost of the tender. Purchase was completed just before VAT was introduced thus a considerable saving was effected.

Above:

End of March 1973. "KE1" has escaped and is on her way! It was a time of triumph and relief for the members. At the time a restoration period of between 3 and 5 years was contemplated.

[6024 Society Collection]

An added problem for the society was that, although blessed with a highly prestigious engine, it is very heavy. Very few preserved railways can run it on their lines so it could not feature in their plans. This is because most preserved railways are branch lines, unwanted and closed by B.R. in the 1960s and never built to carry the heaviest engines.

Negotiations took place with a prestigious railway centre but they became protracted and after several months they were abandoned and eventually "Quainton Road Station" was decided upon which is, of course, now the home of the "Buckinghamshire Railway Centre" near Aylesbury.

An early member, Rod Thomas, has described how several weekend visits were made by members to Barry to prepare the engine for transport and also to purchase spares from No 6023 which it was thought would eventually be scrapped because of its condition.

The "King Preservation Society" had by this time begun to produce a Newsletter called "The Messenger" later known as "The Kings Messenger" (referred to as "KM" from now). Edition No 1 of September 1972 contained a letter from Mr Tankard which started off with these words:

"The King Preservation Society is now 6 months old .... we are now almost ready to announce the purchase of the locomotive from Woodham Brothers' scrapyard in Barry .... we have estimated that the restoration will take between 5 and 7 years dependent upon the size of the restoration team .... on other fronts September has seen the Society represented at open days

The engine was eventually removed from Barry and arrived at Quainton Road on the 1st April 1973 having taken 4 days for the journey. It was carried on a large low loader truck and trailer and the tender was taken separately on a similar vehicle. The engine and tender were placed in a siding on the "Up" side of the museum. No shed or shelter was available.

It is recorded that a number of punctures were suffered by the low loader whilst carrying the engine from Barry and delay was caused in that way and the engine moreover caught on some telegraph wires when nearing its destination!

8: Restoration

Over the 15 months between April 1973 and June 1974 the engine was dismantled.

On the 27th July 1974 the boiler was lifted from the frames and thereafter the frames were lifted from the bogie and the bogie lifted from its wheels; the tender was also removed from its wheels that day.

These operations were amongst the largest carried out on a preserved steam engine at the time. They involved two cranes.

Between 1973 and 1980 volunteers worked in all weathers in the open. Until 1980, the workshop, such as it was, consisted of a medium sized goods van!

The volunteer working parties erected a shelter constructed of scaffolding, tarpaulins and plastic sheets. Many of the locomotive parts, materials and tools were kept in this old Southern Region four-wheeled "PMV" utility van which served us well for a number of years.

The outdoor phase of restoration continued until July 1986 when a shed adjacent to its former siding was completed for the engine and tender. This transformed working conditions which had previously been very hard and trying as can be imagined.

From 1972 to about 1978, working parties of up to about 10 regulars attended weekly at Quainton Road. Thereafter from about 1977 two parties working alternate weekends have run for much of the time.

Until 1977 the early working party was led by Dennis Howells who now owns No 9466. He is still a member of the society and has given considerable and invaluable assistance.

On Dennis' departure another member, Darryl Polley, took up the leadership of the working parties and he continued his work on the engine until 1985.

Under the auspices of Dennis Howells considerable progress was made scraping down and cleaning the frames, wheels and other parts and in obtaining parts which would be needed

Below:
A photograph probably taken in 1973. 6024 has arrived at the up yard in Quainton, where she is pictured next to the boiler of 34016 "Bodmin". The cladding has been removed from the boiler and the splashers and fire-iron cover next to the firebox have been lifted. *[6024 Society Collection]*

Five photographs of dismantling all taken on 27.7.74.
Right: **the cab roof is lifted clear.**
*Below right:***the boiler follows, it will be 13.5 years before it returns to this position.**
Bottom right: **the complete frames, cabsides and footplate clear the wheels.**
Far right: **thereafter the tender wheelsets were rolled free. This part of the restoration of the locomotive on its own represented a considerable hurdle to the Society, both in terms of expense and working hours.**
Below far right: **the tender is lifted. [All 6024 Society Collection]**

in the future and having parts manufactured.

During the period from 1974 to 1977 though, money was short.

Darryl Polley has related that there was little money for any part of the restoration project when he started in the working party. The society had two unsuccessful fund raising appeals.

Following these appeals there was an unsuccessful proposal to issue 25 shares of £1,000 each in the locomotive. It can easily be imagined the stage the project had reached in about 1977 that mainline running hardly seemed around the corner! No doubt this caused many people to search their consciences and hold back from spending large sums of money at this stage. The engine was in pieces, it needed some new rods and other parts, work was proceeding out in the open and many of the other parts which have since been found or manufactured such as atomizers and cab fittings were years away from being acquired. Shares have in fact been purchased in the locomotive since that date, indeed some members have been extremely helpful and forthcoming in this way.

The "100 Club" was agreed on after Ray Freeman became our Chairman. This was a scheme whereby 100 people would contribute £5 per month to a fund and a draw took place periodically for a large prize.

This was far better supported, probably because it gave the possibility of a return to the members over and above the satisfaction of restoration progress. It gave the promise of a steady income and enabled expensive items such as eccentric straps, connecting rods and tube plate to be purchased.

The "King's Messenger" magazines for the 1970s and early years of the 1980s also contained numerous descriptions of members' efforts to raise money by collecting unwanted items and going on sponsored walks. On the other side of the coin there were periodic appeals for money – there never seemed to be quite enough to go round, although one fund raising idea well supported was a car raffle yielding over £3,000 after a great deal of effort.

Throughout the roughly 16 year

restoration period of the locomotive, a steady source of funds has been the sale of books and other railwayana. For many years until 1989 this has been handled by John Twydle, who with his assistants has run the 6024 Sales Coach at Quainton. This is a coach body of great antiquity which John has taken over, refurbished and kitted out as a shop. Not only has he had his opening days throughout the summer season when the Railway Centre has been in operation, but additionally he has taken his sales on the road to various exhibitions including such events as "Woking 150" in 1988 and similar affairs in order to make money for the Society.

Yet a further source has been a regular stream of generous public support in the form of donations in our collecting box' On some Saturdays in the summer in recent years, we have had between 50 and 100 visitors in an afternoon and many of these have given generously. For these donations we have been very grateful.

Last, but by no means least, has been the instrument which has, in recent years, been the mainstay of our treasurer's peace of mind. This has been the arrangement known as "Club 60-24". Those who join covenant to pay a minimum of £8 per month over a period, in return for which they would receive various items associated with the locomotive including a main line trip and secure various privileges as well. This arrangement has been a superb foundation for financing the later stages of the restoration and again we must give grateful thanks to all those who have given covenanted funds in this way.

It is obvious that societies which have purchased locomotives and carried out restorations since ours have learned much from us and our contemporaries. It is of paramount importance to have regular working parties and a steady flow of funds.

Share schemes or covenant contribution schemes with various benefits such as pictures and a guarantee of access to the locomotive and concern in future rides seems to have been the way forward in many cases.

In 1982 Darryl Polley wrote that: ".... for many years little seemed to happen .... the question always was 'Have you got your rods yet?' and the answer was always 'No'. The rods were always our main problem, the prime factor between us and steaming our King .... In April we ordered the forgings for the connecting rods .... the next part is to get the forgings machined."

The final bill for the connecting rods was £4,000 and the machining cost a further £12,000.

The other problem was the smokebox tube plate which was eventually fabricated at Hugh Phillips Engineering in Tredegar at a cost of over £2,000, the original being found to be cracked on internal inspection. After purchase, it lay unfitted for several years until on the 11th February 1987 it and the boiler were taken by road to "Sail and Steam Engineering" at Brightlingsea, Essex. During the overhaul of the boiler, this new tube plate was rivetted in place.

Between 1975 and 1979 work concentrated on the tender, out in the open. It had suffered from its 10 years outside at Barry, and some of the steel platework needed major attention. Wear and corrosion meant that several patches had to be welded into the coal space and onto the tender top (and to its front end, later, in 1987).

The drain tubes to the rear top area had corroded away and new tubes had to be welded in.

The top lamp bracket had to be rewelded on, and the steel pipework was removed and corroded areas replaced, before refitting the vacuum brake system, water feed and steam heat systems. The vacuum reservoir too was extensively corroded and was therefore repaired before repainting and refitting. New steel work was made to hold the fire irons.

Incidentally, it is convenient to mention the many great lengths of steel pipework on the locomotive itself: this was removed and retained for similar treatment as the tender pipes later in the restoration.

For both the tender and engine, new brake blocks were required. Unfortunately, after the tender blocks were made, we found that they were modelled on already worn blocks, so they had to be cast again.

In 1979, on the day the tender was rewheeled, the bearings were scraped in; the remetalled axle-boxes being lifted on and off the skimmed journals by hand.

The tender wheelsets had been sent

to Swindon for the skimming of the journals and profiling of the wheels; in 1987 the wheels were reprofiled to B.R. current requirements. Whilst the wheelsets were away for a second time, the opportunity was taken to repair the front and rear tender dragboxes. New steelwork was fitted to replace complete areas, and new oak planking supports between tender frames and tank were fitted.

The water scoop, although working, has been removed. It is impossible to use it and remains a risk, although very slight, that it could foul some object on the track. Its absence makes the tender underside more accessible too.

Although the working party broke the back of the tender work out in the open, the locomotive took huge strides before the shed was available in July 1986.

One of the most spectacular achievements in the restoration of "6024" has been the restoration, remaking and rebuilding of the motion rods and parts.

The original pistons and valves were retained, but the locomotive has received 4 new piston rods for the pistons and new valve and piston rings throughout. The ports to the cylinders have been decarbonised; after years of running since this was last done before 1962, the carbon deposits in these areas were considerable.

The restored "6024" carries slidebars which spent over 10 years outside at Barry carrying only accumulated oil and dirt. Open to the sea air, they were very pitted by the time they reached Quainton. One of the earliest jobs on the locomotive was to have these parts skimmed at Swindon to remove rust pitting which had thus formed.

One of our long-serving members has re-aligned these slidebars correctly, having become expert in this work.

The eccentric sheaves on the inside motion were machined to clean the surfaces and thereafter the eccentric straps were remetalled by casting white metal into them and machining to size. The die blocks received new white metal and the expansion links were re-machined at Swindon works.

After this work had been done the valve gear was reassembled.

However, we were unhappy with the play in the rocker arm bushes; new brass bushes and washers were turned, so that this assembly was returned to "as new" condition.

The original coupling rods have been skimmed to remove rust, and refitted, but the same could not be done with the connecting rods. All 4 connecting rods, along with the piston rods, had been cut through with oxy-acetylene torches to make the towing of the engine easier. Photographs of the engine at Barry demonstrate the lack of connecting rods. There was no short cut. This was a massive obstacle to full restoration, and although manufacture of rods abroad was mooted in the early days, this was rejected.

Instead the members held on to their avowed course of restoration to B.R. mainline standards, and for this they had to comply with B.R. metal quality criteria.

After one disastrous effort by a contractor to produce the 4 rods, they were eventually forged and manufactured

Above right:

In May 1983 the interior of the Society's workshop at Buckinghamshire Railway Centre is shown, along with some of the essential equipment. There is a generator under the bench to provide electricity and along with the obvious bits and pieces on the shelf above the bench is a box marked "earmuffs" and "emery cloth".

[D. Polley]

Right:

Probably a photograph from the early 1980s which shows work in progress outside on the two live steam injectors. One was an original whose internal parts were missing and the other was manufactured in place of the original 12mm exhaust injector which went missing at Barry.

[M. Wedgebury]

and delivered between 1982 and 1986, giving the members of "Club 60-24" triumphal evidence of the value of their steady contributions to the restoration.

Between these rods and piston rods are, of course, the crossheads; these have been remetalled and refitted with working party members spending many hours at the workbench and between and beside the frames carefully shaping their bearing surfaces.

As to the bores of the cylinders and valves, these have also tested the ingenuity of the working party. After exposure to the sea air, the bores were pitted, so that while the B.R. Inspector pronounced them fit for further use, the surfaces required honing to smooth steel. This was done with a hand machine made up by society members which bolted across the front and rear studs to the cylinder covers, and rotated by hand using large sheets of emery cloth to remove almost all the pitting, after many hours of work.

Above:
Some examples of the less conspicuous works which faces most preservation societies; steam engines are festooned with great lengths of ferrous and copper pipes for vacuum brake, steam heating, blower water feed, ejector and lubrication systems. In many ex-Barry locos the pipes needed wholesale replacement. *[M. Wedgebury]*

Above left:
The eccentric straps are in place and the new connecting rods are awaited in this photograph from the early 1980s. It was taken when the engine was kept outside and covered by large tarpaulins. *[M. Wedgebury]*

Left:
A 1982 photo: Steve Underhill prepares main bearings for refitting. *[Society Collection]*

One of the main attractions of large steam locomotives for scrapmen was the large amount of non-ferrous metal they carried. Apart from the copper fireboxes, which were left in the engines to survive at Barry, brass castings and copper pipework of various bores were removed as most was easily accessible. Woodhams also removed the various gauges in the cab. The copper cap to the chimney survived but the whistle and brass safety valve cover were taken off the engine.

One member kindly donated the brass front bogie axlebox covers, the originals having disappeared.

The society has been fortunate in acquiring replacements for the missing gauges in the cab, which are so much part of the impressive spectacle of a working engine.

On the other hand, many parts were simply unavailable from any known source, and the cost of replacements has gobbled up money at a tremendous rate. The 3 atomizers alone cost around £3,000 to cast in phosphor-bronze.

The cylinders required new pressure relief valves, and oilboxes, and all 11 draincocks had to be manufactured.

New windowframes were cast, fettled and polished, and the steam fountain and driver's brake valve in the cab were newlymade.

The society has been very fortunate to have good friends at the West Midlands firm of Wednesbury Tube. They have generously supplied, free of charge, the copper pipework for the steam heat and water feed to boiler piperuns in the cab, and the blower pipe, and runs to the ejector; and most visibly, the extensive web of small bore lubrication pipework: all was missing on purchase from Barry.

Certain working party members have made it their business to see that these replacements were expertly fitted and ensured their proper functioning: Two in particular have given hours of weekend work to the draincocks and associated parts, and another member has worked tirelessly on the pipe runs for the oil distribution.

Another huge expense has been the injectors. After withdrawal and scrapping, the engine lost both its (right hand) live steam injector, and (left hand) exhaust injector. The society acquired the body to one 10mm live steam injector, minus internal parts: these have been made, together with a further totally new live steam injector. Therefore, as she now runs, "6024" no longer has an exhaust injector.

Some of the most visible brass parts are the splasher beadings: we were fortunate to have 3 of these, but the other 3 were made by one of our members.

As purchased from Barry, the locomotive's wheels were in no state to run straight away. We are, of course, fortunate to have had an engine with its wheels intact, unlike "6023", but they

Above left:
Here is a photograph taken in about 1984 of one of the newly-manufactured outside connecting rods for 6024. The work was undertaken at Dinsdale Engineering Ltd in London. The purchase of these new rods was the most expensive item after the boiler overhaul. *[Society Collection]*

Left:
A smart-looking tender No 2839 is re-wheeled at Quainton after 5 years. Unfortunately the tender wheelsets required further profiling. The drag boxes were partly replaced in 1987 while the wheelsets were away. Also the wooden seating between the frames and the tender tank was renewed. *[Society Collection]*

Above:
By the time this photograph was taken in April 1983, the manufacture of the loco's new front tube plate was almost complete. The 32 larger holes will carry the larger flue tubes within which lie the superheater elements. *[Society Collection]*

Above right:
The frames are shown here in 1984 or 1985 restored and repainted; at the foot of the picture is the vacuum brake cylinder. This is to be fitted under the cab floor. *[M. Wedgebury]*

Right:
On 20th March 1983 the Society hired two cranes. These were positioned at either end of the frames which were lifted off the ground to enable the driving wheelsets and the bogie to be rolled into position. The boiler can be seen on the extreme right. *[Society Collection]*

Above:

On 20th March 1983 the wheelsets are all in position and the frames are lowered to be re-united with them. One clearly visible feature of the bogie is the two bearer plates, one on each of the side frames between the front and rear wheelsets. Plates attached to the main frame rest on these.
[Society Collection]

nevertheless received considerable attention during preservation.

The axle journals were skimmed at Swindon Works and the driving wheels and bogie wheels were reprofiled. The driving wheel tyres will possibly not last for a great deal longer: the society is putting aside money to pay for the considerable work this will entail at an estimated cost of £10,000 £12,000; proceeds from the sales of this book included!

The axleboxes of the bogie and driving wheels were removed at Barry for their brass and white metal content. The society had new wooden patterns made from which replacements were made: new brass castings were completed in the workshop of the Severn Valley Railway. This enabled the rewheeling to take place: a great day in the restoration.

The rewheeling was a delicate operation indeed. Two large cranes were hired and a team standing at the wheelsets ensured the axleboxes entered the horns together as the frames were lowered. The Severn Valley Railway sent two engineers to Quainton beforehand to measure the axlebox horns to ensure that, on rewheeling the locomotive, the centre lines of the axles would remain exactly aligned.

The next day, the event was recorded in "the Daily Telegraph".

Although some of the cab gauges and fittings were purchased early in the restoration when the opportunity presented itself, as indicated, we were not so fortunate with other items. In 1984, we were able to avail ourselves of the visit by "6000" "King George V" to Swindon works for overhaul, as an opportunity to examine parts we were still in need of, and take measurements.

Many of these were backhead controls: when the engine arrived at Quainton, the backhead had no controls, the backhead cladding too has been completely made from scratch.

In the 1970s less progress was made on the boiler than other areas. The boiler on its own has been a huge expense.

The smokebox was removed from the boiler while the engine was at Quainton: the rivets attaching it were cut through, a cracked front tubeplate was discovered, and a replacement was one of the earlier jobs on the boiler.

All the tubes were removed: the large flue tubes were retained with the superheater elements as they were suitable for re-use, with the superheater. The small tubes were discarded and we were fortunate to obtain a new set of 115 small tubes at a very reasonable price.

By Spring 1982, Darryl Polley was able to report in the magazine "... the boiler is the scene of most activity, with rivet and stay removal proceeding at a great pace. The water space around the firebox has been descaled, and in the near future a firm of grit blasters will be employed to descale the inside of the boiler and the outside of the large tubes ... a B.R. boiler inspection is to be arranged in the near future

By July 1984, it was hoped we might attend the Swindon "GWR 150" celebrations in the following year: in the event we were not ready and the event was cancelled, but much effort was given to early completion of the connecting rods: that work was put out to tender in November 1983.

Our hopes of early completion gave added impetus to the boiler work, much of it a considerable challenge, during 1984 to 1986. Much of the society's effort was directed at the firebox.

Right:
A photograph taken in 1984 showing new stay work on the outside of the firebox. Replacement of stays on this side is almost complete. *[M. Wedgebury]*

Below right:
A photograph taken in about 1984 showing an air-driven power tool being used to ream out an old stay-hole. *[Society Collection]*

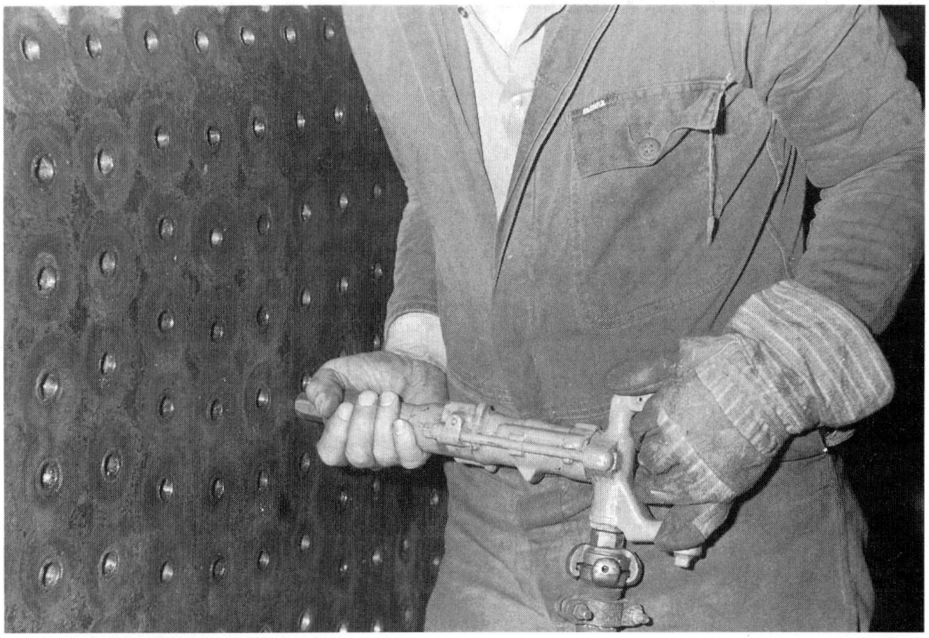

Bottom right:
Many of the firebox stays below frame level were renewed over 800 in all. This picture from about 1984 shows the firebox interior looking towards the cab firehole. On the left of the photograph are the holes awaiting new stays. Below the firehole many new stays are in place. The hexagonal heads are removed. Below the frame line all firebox stays were removed by drilling plus oxy-acetylene, replaced and fitted with nuts to protect the new stays from burning – the originals were "un-nutted"; all work was carried out by society volunteers. *[M. Wedgebury]*

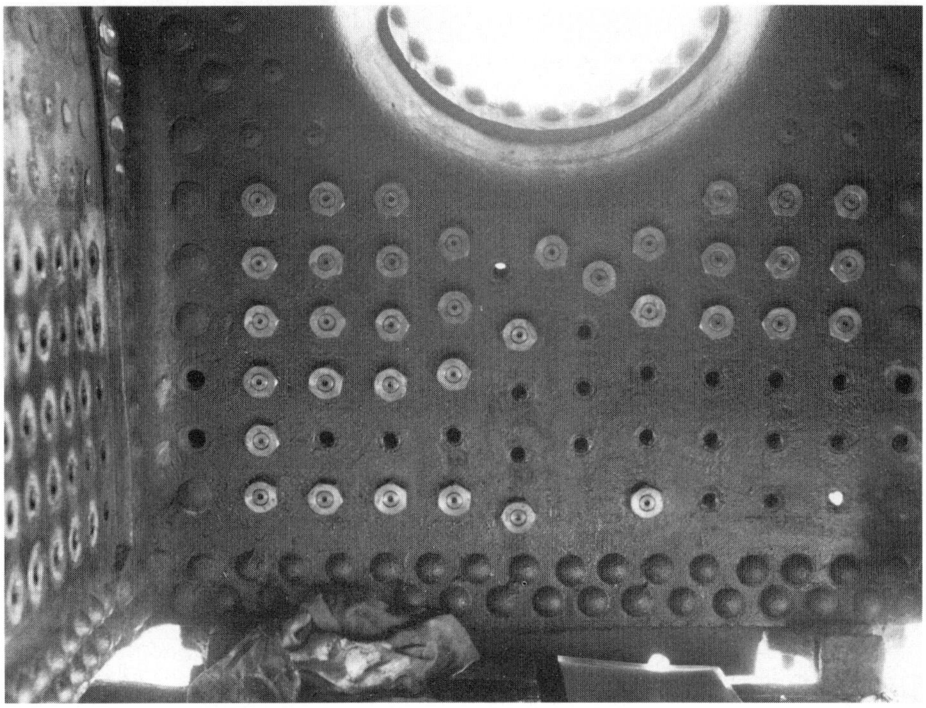

In 1984, the rear flange inside the firebox was completely restudded, after manufacturing special tools and studs to replace back flange rivets up both sides and re-staying began. Approximately two thirds – about 800 – of the firebox stays had to be removed.

The stays are long bolt-like pieces of steel holding the inner copper firebox (copper sheet about 1" thick) to the outer, steel, boiler shell, enabling a gap to exist between inner and outer skins of metal. In this way water can pass between the 2 layers of metal in an area of varying width but often approximately 2"; more at the top. Each skin contains a thread into which the stay is screwed.

The magazine of July 1984 said that "...Restaying is a lengthy procedure which is started initially by burning one side of the stay off and drilling out the other. The stay then falls into the water space. Subsequently this water space has to be cleared of old stays and debris. Each stay hole is then cleared on the steel side and the copper side. Having done this the holes are then reamed to tapping size. They will then be tapped, the stay will be threaded through; one side will be caulked; the stay driver removed, and then the other side caulked; the stay nut fitted on and should the stay protrude, this will be ground off."

In addition the firebox tube plate seam rivets had to be replaced up both sides of the firebox. Rivet snaps and holding up jigs were manufactured and a small air blown forge to heat the rivets and to keep them hot was acquired. Members carried out this work over a two year period. There were many days when more were removed than were finally fitted, but our working members

Above:
February 17th 1987: the boiler departs for "Sail and Steam" at Brightlingsea. *[M. Andrews]*

Right:
A working party!: it is June 1986 and part of the working party is laying track for the new shed. The frames and wheels are on the left.
[Society Collection]

can claim a probable first in locomotive restoration with this work.

Unfortunately, the locomotive did not make the planned deadline of a 1985 steaming because of lack of money and the weight of work. As can be seen above the removal of the old corroded stays was in itself a huge job requiring hundreds of hours patience and perseverence. The committee finally made the decision after failing to make the 1985 deadline to put the boiler out to a contractor in order to speed the restoration programme. This was only possible by soliciting interest free loans from a few members.

Eventually the boiler was taken to Sail and Steam. There some welding was carried out to wasted areas, a new inner water tray beneath the water feed area at the top of the boiler was fitted and a new inner lining welded into the base of the smokebox, the smokebox and tube plate were rivetted to the boiler, new smoke tubes fitted and superheater flues refitted with new forged ends.

The regulator rod and the tubes, superheater elements and superheater were all fitted together with the safety valves.

A great step forward took place in 1986 when the society was able to place the loco undercover for the first time.

The great day in March 1988 came when the B.R. and insurance Inspectors carried out a test of the boiler. Firstly the boiler was filled with water to a hydraulic pressure of 350 lbs per square inch and thereafter steam tested to 250 lbs per square inch and it passed these tests most satisfactorily.

Whilst the boiler was away, more time was spent on the tender frames whilst the wheels were away for the second turning. The tender was jacked up on sleepers and supports and the opportunity was taken to needle-gun the underside again, this work having previously been done in the open in 1981. At the same time partially new drag boxes were fabricated and fitted. This involved extensive welding and fabricating work but it will serve the society in good stead for a number of years to come and was therefore well worthwhile.

On the 24th November 1987 Dennis Howell's locomotive No 9466 was used to tow the newly rewheeled tender and the chassis of the locomotive up and down Quainton Yard. All the bearings throughout the locomotive and the tender had been renewed in some way and the opportunity was taken to find out how the motion was running whilst clear access was available to the inside motion. All went very well apart from some warmth creeping in to the outside big end bearings and one of the cross heads was found to be running slightly askew on the slide bars. The opportunity was taken to attend to these problems thereafter.

One of the most significant days in the restoration work came on March 28th 1988 when the boiler was reunited with the locomotive and the cab roof was refitted having been needlegunned and painted.

Many members witnessed the refitting of the new ash pan to the boiler and thereafter the lowering of the boiler into the frames at Quainton.

A photograph of this event made its way into the pages of "The Daily Telegraph" the following morning! In addition the event was covered by television.

At that stage it looked as if work was going to rush to a speedy conclusion but the working parties knew very well that this would not be the case. Over the Summer and Autumn of 1988 much associated pipework and other fitting work was carried out to the boiler and the new cladding was prepared for fitting. All new backhead cladding was made for the rear end of the boiler which protrudes into the cab and a new cab floor was fabricated and fitted in the Autumn of 1988. At the same time the damper rods were formed and fitted. By this time it was known that the first steaming date would not be far away.

On the 2nd February 1989 the locomotive was eventually ready for its first steaming for 26 years. The fire was lit late in the evening of the 1st February but by 8 a.m. a problem had arisen. The fire had to be dropped whilst modifications were made to a part of the vacuum braking system before steaming could proceed. by an intensive effort the problem was rectified and the fire was re-lit in the morning of the 2nd February. By lunchtime, boiler pressure had reached over 100 lb per square inch. There were some steam leaks but a moving test was felt justified and in the afternoon members were privileged to run the engine on a short section of track in Quainton "Up" yard.

Some members who had been working on the locomotive for over 10 years were present and could not have hoped for a more satisfying and pleasurable outcome to their work.

During the rest of February, the opportunity was taken to cure certain steam leaks and attend to various other minor matters which had been disclosed by the initial steaming and a repeat steaming and running took place on the 5th March.

Above:
"KE1"'s boiler weighs about 35 tons. It received numerous new parts, including 115 new small tubes, and the new front tube plate during repair, but still needed many hours of work. Amongst the tasks to come were the fitting of the backhead, gauges and other equipment and the pipework to complete the exterior. In the smokebox, the steam pipes and spark arrestor cage are yet to be added. *[M. Andrews]*

Left:
2nd February 1989. One of the biggest days in the Society's existence, on 1st February 1989 the fire was lit and by 1 p.m. 6024's boiler pressure had reached the point where she could move under her own steam again. *[Author's Collection]*

Below:
Later on 2nd February 1989, 6024 steamed successfully on the up yard running line at Quainton Road. *[M. Andrews]*

9: "King Edward I" restored

Above:
6024 "KE1" stands at Quainton Road up yard platforms at Buckinghamshire Railway Centre; about two hours before, H.R.H. The Duke of Gloucester had re-named her. *[R.C. Riley]*

Above right:
No 6024 is just north of Henley-in-Arden station with the 18.50 Tyseley–Stratford of 15 April 1990. Note the Birmingham Railway Museum's two GWR saloon coaches behind the engine.

Right:
Storming up Llanvihangel bank with the Newport–Hereford leg of the charter train of 24 June 1990. *[Both: Peter Nash]*

By February 1989 the membership knew that a renaming ceremony had been arranged over the May Day Bank Holiday starting on the 26th April 1989 and work pressed on to that end.

A member of the staff of the Severn Valley Railway was extremely generous in carrying out the insulation and cladding work on the boiler over the weekend of the 18th, 19th and 20th of March, free of charge and we are all greatly indebted to him. Work was carried out with the assistance of working party members under the leaders of the two working parties of several years past.

It should be recorded here that this is by no means the only assistance and guidance we have received from the Severn Valley Railway and Birmingham Railway Museum at Tyseley and the society is greatly indebted to them for the help and guidance over the years of restoration.

We must also record considerable thanks to the Great Western Society at Didcot and the "Duke of Gloucester" Preservation Trust who have given considerable guidance, and other bodies of railway preservation individuals who have given generously of their time and advice.

Before that the loco had a busy season of display and running in the "Up" yard at Quainton on visitors days, following the formal "re-naming" and recommissioning by H.R.H. Duke of Gloucester on April 26th and we were given much press and television coverage, plus mention in the "Court Circular". The society has had considerable support from the Buckinghamshire Railway Centre during the long restoration and it was only right that they should have first opportunity to benefit from public display and running of the locomotive. Unfortunately, on Sunday the 26th May 1989, whilst running, a problem in the machining and alignment of a piston rod and cross head became apparent. Three months work saw rectification and completion of this together with lining and lettering work.

In fact, the locomotive left Quainton on Monday 9th October 1989 to be transported to Tyseley Railway Museum near Tyseley Station in Birmingham. The engine remained there for an initial period of 6 months. Most recently "6024" has taken up residence at the G.W.S. Didcot Railway Centre.

There is a challenge on the horizon. It is the prospect of having to re-tyre the locomotive's main driving wheels in the not too distant future. At one time it was feared that this would have to be done before the locomotive turned a wheel on the main line again but that proved not to be the case. The work will have to be carried out some time during 1990 or early 1991 though and the society is running an appeal to raise funds for the work.

Despite this outstanding item of work, "King Edward I" is in many respects a new engine and this point should not be overlooked when considering the real value of the restoration and overhaul it has undergone at Quainton Road. Of course, any engine which in the past underwent a heavy general repair would have had many of the items seen to which the society has attended to, but as can be seen from the foregoing catalogue of replacement of parts, a great number were simply not there and we were forced to renew. Some of the cab has been renewed as well and in other areas rusted and corroded pieces of steel have been cut away and replaced with a very high standard of finish.

At the time of writing, the locomotive still has 5 years of its 7 year main line boiler certificate to run and we hope we will be able to take considerable advantage of this. Subject to B.R. consents to future arrangements for main line running, not only will you be able to take a seat in a train behind the locomotive on former Great Western Railway lines, but also view it on one of those runs or at a Railway Centre in the

6024

6024

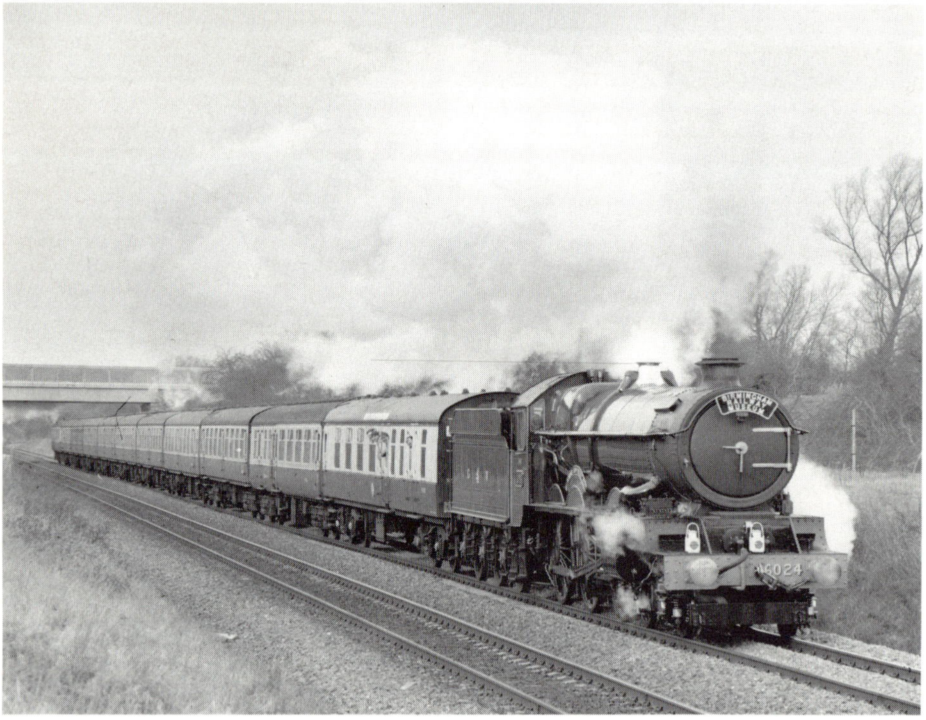

Above:
"KE1"'s first public appearance away from Buckinghamshire Railway Centre was at the "Castles and Kings" weekend at Birmingham Railway Museum, Tyseley. Here is the engine with Nos 5080 "Defiant" and 7029 "Clun Castle" during the weekend in steam. *[R. Stuart-Smith]*

Left:
On 1st February 1990. 6024 "KE1" is seen running at speed at Fenny Compton, south of Leamington Spa on the Birmingham to Banbury former G.W.R. main line. *[R. Blenkinsop]*

Above right:
6024 arrives at Derby with the 'Royal Venturer' charter train which it has hauled from Didcot on 24 November 1990. *[L. A. Nixon]*

Right:
A photograph showing the cab of 6024 in recently restored condition. *[M. Andrews]*

company of other locomotives which may either have been purchased from B.R. or like ours rescued from Barry Scrapyard.

It was the pinnacle of the society's achievements so far, when, in February 1990, the engine ran on the main line for the first time since 1962.

On Tuesday 30th January 1990 the loco ran to Derby without problems, and on the Wednesday it underwent weighing and adjustment to springs.

On Thursday February 1st 1990 the loco backed on to 10 coaches and ran from Derby to Tyseley, then on to Banbury. An inspection revealed that all was well save failed lubrication feeds to the slide bars, but the engine was able to take the train back to Tyseley.

Since the trial run, 6024 has fulfilled all the hopes of the society and surpassed them. Over the weekend of 15th and 16th April 1990, 6024 ran the "Shakespeare Express" excursions from Tyseley to Stratford on Avon, performing superbly with No 5080 "Defiant" of Birmingham Railway Museum.

On Saturday 19th May 1990, "King Edward I" ran from Didcot to Stratford